W9-CCC-891

GENESIS

ALSO BY EDWARD O. WILSON

The Origins of Creativity (2017)

Half-Earth: Our Planet's Fight for Life (2016)

A Window on Eternity:
A Biologist's Walk Through Gorongosa National Park (2014)

The Meaning of Human Existence (2014)

Letters to a Young Scientist (2013)

The Social Conquest of Earth (2012)

Why We Are Here: Mobile and the Spirit of a Southern City,
with Alex Harris (2012)

The Leafcutter Ants: Civilization by Instinct,
with Bert Hölldobler (2011)

Kingdom of Ants: José Celestino Mutis and the Dawn of
Natural History in the New World,
with José M. Gómez Durán (2010)

Anthill: A Novel (2010)

The Superorganism:
The Beauty, Elegance, and Strangeness of Insect Societies,
with Bert Hölldobler (2009)

The Creation: An Appeal to Save Life on Earth (2006)

Nature Revealed: Selected Writings, 1949–2006 (2006)

From So Simple a Beginning: The Four Great Books of Darwin,
edited with introductions (2005)

Pheidole in the New World: A Hyperdiverse Ant Genus (2003)

The Future of Life (2002)

Biological Diversity: The Oldest Human Heritage (1999)

Consilience: The Unity of Knowledge (1998)

In Search of Nature (1996)

Journey to the Ants: A Story of Scientific Exploration,
with Bert Hölldobler (1994)

Naturalist (1994); new edition, 2006

The Diversity of Life (1992)

The Ants, with Bert Hölldobler (1990);
PULITZER PRIZE, GENERAL NONFICTION, 1991

*Success and Dominance in Ecosystems:
The Case of the Social Insects* (1990)

Biophilia (1984)

Promethean Fire: Reflections on the Origin of the Mind,
with Charles J. Lumsden (1983)

Genes, Mind, and Culture, with Charles J. Lumsden (1981)

On Human Nature (1978);
PULITZER PRIZE, GENERAL NONFICTION, 1979

Caste and Ecology in the Social Insects,
with George F. Oster (1978)

Sociobiology: The New Synthesis (1975); new edition, 2000

A Primer of Population Biology, with William H. Bossert (1971)

The Insect Societies (1971)

The Theory of Island Biogeography,
with Robert H. MacArthur (1967); reprinted, 2001

GENESIS

The Deep Origin of Societies

EDWARD O. WILSON

Illustrated by Debby Cotter Kaspari

LIVERIGHT PUBLISHING CORPORATION

A DIVISION OF W. W. NORTON & COMPANY

Independent Publishers Since 1923

New York London

Printed in the United States of America
First Edition

For information about permission to reproduce selections from this book, write to Permissions, Liveright Publishing Corporation, a division of W. W. Norton & Company, Inc., 500 Fifth Avenue, New York, NY 10110

For information about special discounts for bulk purchases, please contact W. W. Norton Special Sales at specialsales@wwnorton.com or 800-233-4830

Manufacturing by LSC Communications, Harrisonburg
Book design by Lovedog Studio
Production manager: Anna Oler

Library of Congress Cataloging-in-Publication Data

Names: Wilson, Edward O., author. | Kaspari, Debby Cotter, illustrator.
Title: Genesis : the deep origin of societies / Edward O. Wilson ; illustrated by Debby Cotter Kaspari.
Other titles: Deep origin of societies
Description: New York : Liveright Publishing Corporation, a division of W. W. Norton & Company, [2019] | Includes bibliographical references and index.
Identifiers: LCCN 2018050101 | ISBN 9781631495540 (hardcover)
Subjects: LCSH: Animal behavior. | Behavior evolution. | Behavior genetics.
Classification: LCC QL751 .W55 2019 | DDC 591.5—dc23
LC record available at https://lccn.loc.gov/2018050101

Liveright Publishing Corporation, 500 Fifth Avenue, New York, N.Y. 10110
www.wwnorton.com

W. W. Norton & Company Ltd., 15 Carlisle Street, London W1D 3BS

1 2 3 4 5 6 7 8 9 0

CONTENTS

PROLOGUE

ALL QUESTIONS OF PHILOSOPHY THAT ADDRESS the human condition come down to three: what are we, what created us, and what do we wish ultimately to become. The all-important answer to the third question, the destiny we seek, requires an accurate answer to the first two. By and large, philosophers have lacked confirmable answers to the first two questions, which concern the deep prehuman and human past, thereby remaining unable to answer the third question, which addresses the human future.

As I now approach the end of a long career studying the biology of social behavior in animals and humans, I've come better to understand why these existential questions defy introspection by even the wisest of thinkers, and, more importantly, why they have been so easily enslaved by religious and political dogma. The principal reason is that while science and its attendant technology have grown exponentially, with a doubling time of one to several decades according to discipline, they have only recently begun to address the meaning of human existence in an objective and persuasive manner.

For most of history, organized religions have claimed sovereignty over the meaning of human existence. For their founders and leaders the enigma has been relatively easy to solve. The gods put us on Earth, then they told us how to behave.

Why should people around the world continue to believe one fantasy over another out of the more than four thousand that exist on Earth? The answer is tribalism, and, as I will show, tribalism is one consequence of the way humanity originated. Each of the organized or otherwise public religions as well as scores of religion-like ideologies defines a tribe, a tightly knit group of people joined by a particular story. The history and moral lessons it contains, often colorful, even bizarre in content, are accepted as basically unalterable and, more importantly, superior to all competing stories. The members of the tribe are inspired by the special status the story gives them, not just on this planet but on all other of the multitude of planets in each of the trillion galaxies estimated to compose the known universe.

And best of all, cosmic faith is the bargain price asked for guaranteed personal immortality.

In *The Descent of Man* (1871), Charles Darwin brought the whole subject into the purview of science by suggesting that humanity descended from African apes. Shocking as that was at the time, and still unacceptable to many, the hypothesis has nonetheless proved correct. An understanding of how the great transition from ape to human occurred has been steadily improved since, chiefly by a consortium of researchers in five modern disciplines: paleontology, anthropology, psychology, evolutionary biology, and neuroscience. As a result of the combined labors of scholars in these disciplines we have today an increasingly clear picture of the real creation story. We know a good deal about how humanity was born, and when, and how.

This factual story of the creation has turned out to be vastly different from that first believed not just by theologians but also by most scientists and philosophers. It fits the evolutionary histories of other, nonhuman lines, of which seventeen have so far been found to possess advanced societies based on altruism and cooperation. These are the subjects of the sections immediately to follow.

In later pages I'll take up a closely related subject, also in an early stage of investigation by scientists. What was the force that made us? What, exactly, replaced the gods? This remains a source of contention among scientists, which I will try fully and fairly to address.

GENESIS

I

THE SEARCH FOR GENESIS

THE KEY TO THE LONG-TERM SURVIVAL OF HUMANity depends on full and correct self-understanding, not just of the past three thousand years of literate history, not across the ten thousand years of civilization begun during the Neolithic revolution, but back two hundred thousand years, with the emergence of fully formed *Homo sapiens*. And still farther back, across millions of years of prehuman lineage. With this self-understanding it should then be possible to answer with confidence the ultimate question of philosophy: What was the force that made us? What replaced the gods of our ancestors?

The following can be posed with near certainty. Every part of the human body and mind has a physical base obedient to the laws of physics and chemistry. And all of it, so far as we can tell by continuing scientific examination, originated through evolution by natural selection.

To continue the basics: evolution consists of a change in the frequency of genes in populations of species. A species is defined (often imperfectly) as a population, or series of populations, whose members freely interbreed or are capable of freely interbreeding under natural conditions.

The unit of genetic evolution is the gene or ensemble of interacting genes. The target of natural selection is the environment, within which selection favors one form of a given gene (called an allele) over other forms (other alleles).

During the biological organization of societies, natural selection has always been multilevel. Except in the case of "superorganisms," as found in a few kinds of ants and termites, where subordinates form a sterile working class, each member competes with other members for rank, mates, and common resources. Natural selection simultaneously operates at the level of the group, affecting how well each group performs in competition against other groups. Whether individuals form groups in the first place, and how, and whether the organization grows more complex, and to what effect—all this depends on the genes of its members and on the environment in which fate has placed them. To understand how the laws of evolution include multilevel selection, first consider what both levels are. Biological evolution is defined generally as any change in the genetic constitution of a population. The population consists of the freely interbreeding members of either an entire species or of a geographic segment of the species. Individuals freely interbreeding under natural conditions are defined as constituting a species. Europeans, Africans, and Asians freely interbreed (when not separated by culture), hence we are all members of the same species. Lions and tigers can be hybridized in captivity, but never did so where they once lived together in the wild in southern Asia. Hence they are considered different species.

Natural selection, the driving force of biological evolution in both individual and group selection, is captured in a single phrase: *mutation proposes, the environment disposes.* Mutations are random changes in the genes of a population. They can occur either, first, by an alteration in the sequence of DNA letters of the genes, or second, by changes in the number of copies of the genes in the chromosomes, or third, by a shift in the location of the genes in the chromosomes. If the traits prescribed by a mutation prove relatively favorable in the surrounding environment to the survival and reproduction of the organism carrying it, the mutant gene will multiply and spread through the population. If on the other hand the traits prove unfavorable in the environment, the mutant gene will remain at a very low frequency or disappear entirely.

Let us imagine an example to explain simply (although no real example is ever completely textbook simple). Start with a population of birds comprising 80 percent with green eyes and 20 percent with red eyes. Green-eyed birds have lower mortality and thereby leave more offspring in the next generation. As a result the next generation of the population of birds has changed to 90 percent green-eyed individuals versus 10 percent red-eyed. Evolution by natural selection has occurred.

To grasp the evolutionary process it is immensely important to answer two inevitable questions in a scientific manner. The first is, for variation in any trait that can be measured, such as size, color, personality, intelligence, and culture, how much is due to heredity and how much is due to environment? There is no either/or that fits each trait. There is instead heritability, which measures the amount of variation in a particular population at a particular time. Eye color has near total heritability. It is correct to say eye color is "hereditary," or "genetic." Skin color on the other hand has high but not total heritability; it depends on genetics but also on the amount of exposure to the sun and sunscreen. Personality and intelligence have middling heritability. A kind and extroverted genius can arise from a poor and uneducated family, and an ill-tempered dunce from wealth and privilege. Education, fitted to the needs and potential of all its members, is the key to a healthy society.

Are there enough genetic (high heritability) differences among human populations to distinguish as races—or, put more technically, subspecies? I bring this subject up because race remains a minefield through which stumble the politically self-serving left and right. The solution to the problem is to walk around the minefield and proceed to rationally more fertile ground. Races are defined as populations, and as a result are almost always arbitrary. Unless the population exists apart and to some degree isolated, it serves little purpose to distinguish races. The reason is that when genetic traits change across the geographic range of a species, they almost always do so discordantly. For example, size can vary north to south, color from east to west, and a diet preference in polka-dot pattern across the whole range of the species. And so forth indefinitely with other genetic

Scientists consider
evolution no longer a
theory but a proven fact.
And natural selection
of random mutations
as the grandmaster
of evolution has been
convincingly demonstrated
through field observation
and experimentation.

traits, until the true pattern of geographical variation is hopelessly divided into a huge number of small "races."

Evolution is always occurring in every population. At one extreme, its pace has been swift enough to create a new species in a single generation. At the opposite extreme, the rate of change has been so slow that the defining traits of the species have remained close to those of distant ancestors of the species. These laggards are informally called "relicts" or "living fossils."

An example of relatively fast evolution was the growth of the hominid brain across a million years, from about 900 cubic centimeters in *Homo habilis* to 1400 cubic centimeters in its descendant *Homo sapiens*. In sharp contrast, species of cycads and crocodiles have changed in most of their traits relatively little during the past one hundred million years. They are correctly called "living fossils."

Let us turn now to another subject of sociobiology of basic importance to understanding the evolution of biological organization. It is phenotypic flexibility, the amount of change in a phenotype (the trait prescribed by a gene) through differences in the environment. The kind and amount of flexibility—since these are also genetic traits—can also evolve. At one extreme, the genes prescribing flexibility can be shaped by natural selection to allow only one trait out of many conceivable, such as one eye color inherited by a particular person. At the opposite extreme, flexibility can also evolve to generate multiple possible responses, each fitted to a particular challenge from the environment.

In this case the phenotypic flexibility still prescribes a rigidly genetic rule, as in *eat fresh food, avoid spoiled food* (unless you are a blowfly or a vulture).

Programmed phenotypic plasticity can be much more subtle than any brief description is able to convey. For example, a species' genes can be altered to prescribe what psychologists call prepared learning, a tendency to learn quickly and respond to particular stimuli more strongly than to other stimuli of similar kind. "Imprinting" is the most familiar form. With one experience the young animal learns a particular appearance or scent out of many competing in the environment, and thereafter responds fully to it alone. Newly hatched goslings attach not just to their mother goose but to the first moving object they encounter after hatching. A newborn antelope fixates on the scent of its mother, and she similarly commits to her offspring's odor. An ant learns the odor of her natal colony within the first several days after eclosion into six-legged maturity and remains allegiant to it for the rest of her life. If captured as an immature pupa by a colony of slave-making ants, she imprints on the alien's odor and attacks her own sisters from the mother colony.

An especially significant paradigm of phenotypic plasticity is provided by the Nile bichir (*Polypterus bichir*), one of the lungfishes that can leave the water and crawl on land. The bichir and others among the world's lungfishes are often cited as close in ancestry to the original species that left the water during the Paleozoic Era over 400 million years ago, subsequently evolving into land-dwelling amphibians. In other words, it is an evolutionary line from one world into another. A recent series of experiments reported by Emily M. Standen at the University of Ottawa and her coworkers has added credibility to this scenario. These researchers confined newly-hatched bichirs on land for eight months, then mixed them with other bichirs raised in water. The land-reared group walked faster and more skillfully than did the water-reared ones. They held their heads higher and undulated their tails less. Even their anatomy changed: the bones of their anterior body region grew in a way that gave the fins greater power to serve as substitute legs.

These and similar examples from living species illustrate how the plastic expression of genes in anatomy and behavior can ease major changes in adaptation—and might well have done so in the case of the principal transitions.

To take this argument further, the multiplication of castes in ants and termites was achieved in evolution by extreme forms of phenotypic plasticity. It was Darwin who made this discovery, and by his own account used it to save the theory of evolution by natural selection. Worker ants, which are highly modified sterile females, nearly defeated the great naturalist. He found them, as reported in *The Origin of Species*, "the special difficulty, which at first appeared to me insuperable, and actually fatal to my whole theory. I allude," he said, "to the neuters or sterile females in insect-communities: for these neuters often differ widely in instinct and in structure from both the males and fertile females, and yet, from being sterile, they cannot propagate their kind."

*The Nile bichir (*Polypterus bichir*), a lungfish able to modify its legs and behavior within a life span to accommodate the land or the water. This example is widely believed to illustrate how the land was originally conquered by vertebrate animals, including our own, very distant ancestors.*

Darwin's solution in *The Origin* presents the first account of the concept of evolution of flexibility of genes. It also introduces the idea of group selection, in which advanced social evolution is driven by the hereditary traits of entire colonies, as opposed to individuals within colonies, which for their part serve as the targets of natural selection:

> *The difficulty, though appearing insuperable, is lessened, or, as I believe, disappears, when it is remembered that selection may be applied to the family, as well as to the individual, and may thus gain the desired end. Thus, a well-flavored vegetable is cooked, and the individual is destroyed; but the horticulturist sows seeds off the same stock, and confidently expects to get nearly the same variety . . . Thus I believe it has been with social insects: a slight modification of structure, or instinct, correlated with the sterile condition of certain members of the community, has been advantageous to the community: consequently the fertile males and females of the same community flourished, and transmitted to their fertile offspring a tendency to produce sterile members having the same modification. And I believe this process has been repeated, until that prodigious amount of difference between the fertile and sterile females of the same species has been produced, which we see in many social insects.*

These two processes, the origin of controlled flexibility in gene expression and group selection, were foreshadowed by Darwin to save his theory of evolution by natural selection. I'll now show how they help make possible our modern understanding of the greatest advances of evolution, including the origin of societies and our place in the world.

2

THE GREAT TRANSITIONS OF EVOLUTION

Earth's biological history began with the spontaneous origin of life. It led across billions of years through the formation of cells, then organs and organisms, and finally, in an episode lasting a relatively mere two to three million years, it created species able to understand what had been going on. Humanity, gifted with an infinitely expanding language and the power of abstract thought, was able to visualize the steps that led to its own origin. Called the "great transitions of evolution," they unfolded as follows.

1. The origin of life
2. The invention of complex ("eukaryotic") cells
3. The invention of sexual reproduction, leading to a controlled system of DNA exchange and the multiplication of species
4. The origin of organisms composed of multiple cells
5. The origin of societies
6. The origin of language

There exist residues of all the great transitions in your body and mine; they carry the products of every step in the history of life. First there was the origin of microbes, represented by modern species of bacteria teeming in our alimentary tract and elsewhere throughout our bodies, ten times more in number than cells that carry our personal DNA. Next are the genetically human cells, whose ancestors were made more complex very early by the fusion of microbial cells followed by their transformation into mitochondria, ribosomes, nuclear membranes, and other components that make possible the efficiency of the present-day cell formations. The cells are called "eukaryotic" to distinguish them from the simple "prokaryotic" cells of bacteria. Next in our personal corporeal history book are our organs, constructed from masses of the eukaryotic cells by jellyfish and sponges and other creatures of the ancient sea. And finally came the human person, programmed to form societies organized by a complex blend of language, instinct, and social experience.

So here we stand, and walk, and when agitated run, having arrived helter-skelter after 3.8 billion years of lineage endowed with no certain purpose beyond carrying still forward the vagaries of mutation and natural selection, erect, bipedal, bone-strutted bags of salty water led by guidance systems engineered back in the Age of Reptiles. Many of the chemicals and molecules circulating in our liquid (by weight 80 percent of the body) are roughly the same as in the primordial sea. Our thought and literature remain energized by the widespread belief that all of prehistory and history, including every great transition, somehow served the purpose of placing us upon the Earth. Everything, it has been argued, from the origin of life 3.8 billion years ago was meant for us. The spread of *Homo sapiens* out of Africa and around the habitable world was somehow preordained. It was meant to establish our rule of the planet with the inalienable right to treat it as we please. That mistake, I suggest, is the true human condition.

So let us look more closely at the great transitions. The first transition and the most difficult to visualize is the origin of life itself. The event has been very broadly and accurately conceived, but a lot of uncertainty remains in the fine detail. The first organisms on Earth, by general assent closely similar to bacteria and bacteria-like Archaea, were self-assembled into replicating systems out of the virtually endless random combinations of molecules present in the primordial sea. The particular habitat of this breakthrough is not known, but present opinion favors underwater volcanic vents. Existing cracks in the ocean floor constantly heat and churn chemical-rich water, as it did in primordial times. Outward from the center of the erupting spumes network occur an abundance of physical and chemical gradients that serve as a natural laboratory for random molecular engineering.

How did it all start? We will have a much better idea of the place and manner of the origin of life when biologists create it, when they take chemical compounds synthesized in the laboratory and construct organisms comparable to those living in the world.

A great deal more will be learned if we find life on other planets, whether in distant star systems or those close to home. The most likely sites in our own solar system include the kilometer-deep aquifers of Mars. Let's drill and see! Perhaps more promising is the ice-encased ocean of Jupiter's moon Europa, made accessible by deep fissures in the surface. Let us drill them through to the liquid water and find out. An engineering feat of this magnitude was recently made by the drilling of a thick ice cap in Antarctica to reach the million-year-old waters of Lake Vostok. An astonishing variety of organisms were found living within it, all awaiting biological research.

Another prime candidate is the liquid water likely to be pooled on the ground around the fiery spumes that continuously explode from Saturn's moon Enceladus. The water immediately vaporizes to enter the ring around Saturn formed by Enceladus, but (perhaps!) not before forming liquid short-lived pools. Within which . . .

Both the creation of artificial organisms and the discovery of extraterrestrial life elsewhere in the solar system would be so stunning in impact, and so far-reaching in potential scientific advance, as to earn the status of the seventh and eighth great transitions on this planet.

The second major evolutionary advance, meanwhile, was the transformation of bacterium-level cells into much more complex eukaryotic cells, of which the human parts of our bodies are made. This step, achieved at about 1.5 billion years ago, was the acquisition of mitochondria, nuclear membranes, ribosomes, and other organelles ("little organs") principally by the capture of some kinds of cells by others. The ensemble of organelles yielded a far more effective division of labor of elements within each cell. And that achievement set the stage for larger, more complex organisms.

The third advance, the invention of sex—the controlled and regular exchange of DNA between cells—produced greater variability in adaptation to the environment. Evolution was thereby equivalently accelerated.

The fourth major transition was the assembly of eukaryotic cells into multicellular organisms. Parallel to the organelles within each cell, the collectivity of cells tightly interlocked and organized into an organism permitted the origin of specialized organs and tissues and by that means provided a far greater range in size and form of living creatures. From the oldest known fossils, we can place the origin of multicellular organisms, including the ancestors of all animal species, at no later than 600 million years before the present.

The fifth transition was the assembly of individual organisms of the same species into groups. The culmination of this new step was the emergence of eusocial groups, defined as the high level of cooperation and division of labor in which some specialists reproduced less than others. In other words, eusocial species are those practicing altruism. The earliest known origins of eusocial colonies occurred in the termites, dating back to the Early Cretaceous Period, about two hundred million years before the present. The termites were followed by the ants roughly fifty million years later, and the two together—the termites consuming dead vegetation and the ants consuming termites and other small prey—thereafter came to dominate the ecology of the insect world. Among the African hominin ancestors of the present-day human species, eusociality was most likely reached—by the ancestral *Homo habilis*—no later than two million years before the present.

Cooperation among individuals in a group can be envisioned as originating and evolving by various forms of interaction. First, there is kin selection, in which action of an individual promotes the survival and production of relatives other than offspring. The closer the degree of kinship (as between siblings compared to cousins), the more effective the influence. Even if the altruist suffers losses, the genes it carries shared with the relative by their common descent are benefitted. Most people are more likely to risk life and fortune to help a brother, for example, than a third cousin. Intuitively viewed, kin selection is most likely to promote favoritism within groups, but there are circumstances in which it could help to originate groups.

A second practice that can favor the origin of cooperation is direct reciprocity, a trade between individuals. Ravens, vervets, and chimpanzees are among the many animals prone to form groups by individuals summoning fellow members to newly discovered food. Individual songbirds "mob" with others of their own and other species to harass and chase away hawks and owls that try to settle nearby.

Among the millions of species around us are survivors, evolutionary products that one way or another reveal the six major steps of evolution leading from single-celled bacteria and other single organisms to humanity's advanced capacity for language, empathy, and cooperation.

Regardless of kinship or individual trade-offs, cooperation can be triggered by indirect reciprocity, the advantage of an individual gained by joining a group just to further its own interests. If you were to separate one starling from a flock, it would proceed to forage for food in mostly the same manner as it did within the flock. Alone, however, it would have much greater difficulty in finding enough food, especially when raising a family. It would run a higher risk from predators when on the hunt by itself. On the other hand, when in a group its chance of flying straight to a rich source of food is greatly improved, if at least one member knows the location, and the group will more likely see an approaching predator, and sound an alarm.

In a relative blink of geological time, our species invented language, bringing in the sixth great transition. By this I mean true language, not facial expression, bodily postures and movements, grunts, sighs, frowns, smiles, laughter, and other paralinguistic signals shared by most humans. And not the creative chatter of parrots and crows, the sweet calling of songbirds, or the howls, roars, and chittering of mammals, no matter how varied and modulated.

Animals can communicate by sound, as we do magnificently well, but they cannot truly speak. True language, uniquely practiced by humans, consists of words and symbols invented and assigned arbitrary meaning, then combined to create an infinite variety of messages. (If you doubt the endless productivity of language, choose one out of the infinite series of prime numbers, then count verbally from

Mobbing in order to neutralize a predator. Birds from a common nesting area gather around an intruding hawk (in the center) and cooperate to drive it away from their nests and young. (The locality is the artist's backyard in Oklahoma. The intruder is a sharp-shinned hawk, and the mob consists of blue jays, Bewick's wren, and a red-breasted nuthatch.)

there.) The messages generate stories, imagined and real, variously from all times past, present, and future.

To speech was added literacy, which rendered every human thought potentially global. Humans could ask any question about all the life around them, species by species, organism by organism. The capacity for language, science, and philosophical thought made us the steward and mind of the biosphere. Can we muster the moral intelligence to fulfill this role?

3

THE GREAT TRANSITIONS DILEMMA AND HOW IT WAS SOLVED

THE MAJOR EVOLUTIONARY TRANSITIONS POSE among them one of the premier questions not only in biology, but in the humanities as well: how can altruism arise by natural selection? In particular, at every transition how was it possible to increase the personal longevity of organisms and their reproduction in competition with other group members without lowering their own fitness? What process of evolution can simultaneously increase the welfare of the group at the expense—sometimes fatal—of its individual group members?

The consequences of the transitions dilemma range throughout biology and the deep history of human social behavior. How are we to explain the heroic full measure of a soldier killed in battle, or a monk's lifetime vow of poverty and abstinence? How the ferocity of self-negating patriotism and religious faith?

The same challenge exists in the growth and reproduction of cells that form an organism. Some of the cells, for example epidermal cells, red corpuscles, and lymphocytes, are programmed to die at a specified time in order to keep the other cells alive. Failure to do so precisely on time and in the right place can cause a disease that puts all the cells at risk. Suppose that just one of the many kinds of cells chooses to reproduce selfishly. Then, acting like a bacterium dropped into a large pot of nutrients, it multiplies out of turn to produce a mass of daughter cells. In other words, it turns into a cancer. Why should any one or all of your other trillions of cells not follow suit? Why, with no sense of the world to which it belongs, does it refrain from acting like a bacterium? That of course is the key practical question of cancer research.

This rule of extreme improbability might well be called the dragon challenge of evolution. The original dragon challenge was built up Tianmen Mountain in China's Hunan Province. It contains 99 hairpin turns followed by 999 45-degree steps, ending in a natural rock arch named Heaven's Gate, the forbidden mythic entrance to the home of the gods. The Tianmen challenge is difficult to climb by foot, especially the almost vertical steps. Yet it has been conquered even by motorcycle and by automobile. And in evolution so has its namesake, at least a half dozen times.

How was the dragon challenge of biological evolution beaten—and in a way that yielded the present-day fauna and flora of Earth? And humanity? A solution to the transitions dilemma may be found within what amounts to a second dilemma, as follows. Evolution by natural selection can proceed rapidly. For example, consider the file of a particular form of a gene, called an allele (number 1), in competition with a second allele (number 2) generation by generation.

In each major transition in evolution, altruism at a lower level of biological organization is needed to reach the one above, as in cell to organism and organism to society. The dilemma, which at first seems paradoxical, is in fact susceptible to explanation by evolution through natural selection.

The origin of groups and mystery of human altruism.

Suppose that the frequency of allele no. 1 is only 10 percent when it acquires an advantage of 10 percent over the second allele. The difference may seem hopelessly small, but within one hundred generations, the fraction of the population carrying allele no. 1 would increase from 10 percent to 90 percent. In short, although natural selection is a potentially very powerful driver of episodic evolutionary change, it has seldom delivered.

The second dilemma is the problem of why, given the potential of natural selection, it has taken the major transitions of evolution so long to occur, mostly in the million- to billion-year range.

Basically the same necessary altruistic restraint has existed at every major transition of evolution. At the level of the origin of society, one selfish ant or termite can weaken and doom its entire colony. A single psychopathic dictator can destroy an entire nation. The potential contest of individual versus group pervades all levels of life, from cells to empires. The conflicts they generate fill the textbooks of the social sciences, and they endlessly enrich the humanities.

Restraint and altruism resist scientific explanation because they seem at first so difficult to achieve by biologically evolving populations. To spread, they must impose at each level of biological organization from cell to society a powerful counterforce of natural selection against the "ordinary" natural selection already in place among the units in the next level of biological organization below. The group, for example, must overcome the regency of the organism and the seeming absolute priority of selfish personal success.

Although the problem attending restraint and altruism at the great evolutionary transitions remains clouded by controversy, and almost no scientific explanation is ever complete and final in every detail, I believe that the big picture is at last coming into focus. In the case of the origin of societies from aggregates of organisms, the problem has been largely solved. The advance in understanding has succeeded through genetic theory applied to experimentation and field research, most of it conducted during the present century.

The solution begins with an appreciation of the enormity of the problem and the improbability, in fact near impossibility, of its solution. The great transitions together, composing the dragon challenge of evolution, lead through a field of extreme difficulty.

Similarly, each of the transitions required almost unimaginably vast numbers of components (chemical compounds to simple living cells to eukaryotic cells and so on up), consuming long geologic periods of time, to create the next higher level.

Each transition required, or at least was enhanced by, multilevel selection—occurrence of natural selection at the group levels added to selection at the individual level. What is the evidence?

4

TRACKING SOCIAL EVOLUTION THROUGH THE AGES

THE MOST EFFECTIVE WAY TO DECIPHER THE birth and subsequent evolution of societies, as for all other biological processes and systems, is to find out what actually happened. This direct approach is made feasible by the existence of tens of thousands of contemporary species that display among them almost every conceivable level of evolving social complexity.

The most elementary organized groups above those of bacterial colonies are mating swarms of insects. They are the phantoms of nature, here one hour and gone the next. Among those most commonly seen are the chironomid midges. Individuals, when flying alone, are almost invisible. Such airborne micro-insects belong to the great assembly of very small flies, parasitic wasps, beetles, aphids, thrips and others that you rarely see unless you pay deliberate attention to the minutiae of nature. When flying singly they are like dust particles carried by wisps of air, visible only when passing close to your eyes. Their existence becomes clear when winged adults of one of the species gather in hundreds or thousands in aerial swarms in order to mate. They dance about like acrobats in tight, roughly spherical groups measuring from under a meter to tens of meters across. Their swarms seem to hang in the air. If you pass your hand through one (don't worry, they don't bite), the group disintegrates into swirling fragments. When you pull your hand back, the group reunites.

Similar mobs of sexual frenzy are formed by flies of many kinds, males and briefly virgin queens of a few species of ants and termites, plus assorted insects from springtails to cicadas and butterflies. According to species, they form living mats spread over bare patches of ground. Or lines and clusters along fallen tree trunks. Or, in still other species, they spiral upward into the treetops and out beyond, into the open sky. The most spectacular to human eyes are the leks of grouse, bustards, and manakins. Grandest of all birds are the thirty-two species of birds of paradise. The performing males gather at the chorus-line cluster called leks, some having traveled from far away, to parade in a struggle for the attention of watching females.

Perhaps on life-bearing planets in another star system (which we may reasonably assume exist, somewhere) mating swarms have evolved into something other than a free-ranging competition for sex, but not here on Earth. One faint exception of which I am aware is the cooperation of brothers in the leks of American turkeys. The pairs strut and preen together and join to bully their competitors out of the arena.

At least a second opening for life's slow evolutionary advance toward greater complexity exists in persistent feeding groups. Flocks of common starlings, for example, often

The vast numbers of species that display different kinds and degrees of social behavior allow scientists to reconstruct the likely steps that led to human and other advanced societies.

fly and feed together. Their murmurations, as the flocks are called, comprise from fewer than a dozen to more than a million individuals, the number depending on their immediately available food supply. The largest flocks darken the sky in gigantic swirling formations. When they land to roost they press together, clothing trees as thickly as the leaves. When they gather to feed, their hordes form dark moving blankets up to acres across the ground. Starlings are specialized predators on grasshoppers and other insects that live in short grass. It is advantageous for individual starlings to share knowledge of the most productive sites. Their strategy is to follow leaders who know the places that consistently harbor large quantities of insects.

In such cooperative labor we find expressed the universal principle of modularity, the tendency of all biological systems to divide one way or another into semi-independent but cooperative groups. Members of the different groups specialize in function, even if just temporarily, in a way that serves the overall assembly as a whole and thereby on average benefits each individual singly.

It is interesting to watch, as I have in the suburban neighborhoods of New England, for modularities in the way a flock of starlings leaves its roost and proceeds to a feeding ground. Some of the dense rows of birds resting on the treetop branches and telephone lines start to grow restless. One or several together then fly up and out and land on another tree or line nearby. These leaders and others close behind them evidently remember a productive feeding site, and proceed to move in the correct direction bit by bit, cautiously. Soon the size of the ambient flock grows. Then it suddenly accelerates. The mass foray unfolds swiftly by positive feedback. The more birds flying out, the more that take off to follow. In minutes the entire flock is airborne.

Once arrived at the feeding site, many of the older and more experienced starlings begin to dig small holes to expose insects on the grass roots and in the soil. The younger, less experienced birds use the excavations to pick up leftover prey. Soon another modularity appears, "rolling," in which birds working near the rear of the feeding flock take off and fly in a wave to the front. The entire flock thus rolls forward, harvesting a continuously fresh supply of insect prey.

The formation of a starling flock confers other advantages on the individual members besides an increased food supply. There is safety in numbers from their enemies—cats, foxes, weasels, and other predators on the ground and hawks that circle above. Working as a feathered thousand-eyed Argus, they turn into a single sprawling sentinel. A sudden flash of wings, anywhere within the flock, even if less than a liftoff, alerts others. Within seconds the entire assembly may take off and swirl high in unison, soon to land elsewhere in a different array.

There is safety in numbers. The mammals and birds that prey on the starlings belong to the adjacent higher link in the food chain. Their populations are correspondingly smaller than those of their social prey, even if the latter remain low. Starlings are further protected by prey saturation. There is a

Mating midges and winged ants (above) thwart predators by emerging simultaneously in large numbers. Starlings also tighten their flocks, making it dangerous for hawks to dive through them.

strict limit on how much any kind of predator can consume, the more so if its own population is thinned by territorial aggression among the members of its species.

Finally, there exists another way the starling flock can protect itself by sheer force of numbers. Whether by accident or design of natural selection, the close formation of the airborne group presents a physical barrier to birds of prey. When a hawk swoops into the flock to strike any one starling selected as prey, it risks crashing into another bird by accident. The problem is one of simple aerodynamics. A peregrine falcon, which dives vertically up to 320 kilometers per hour while twisting its body in order to rake a flying bird with its extended talons, is in special danger. A meal of starlings does not come cheap.

Modularity as the automatic formation of subgroups is a forerunner of cooperation and division of labor. It has been lifted this way to a high level of sophistication even in relatively primitive organisms. Among these ur-societies are some kinds of bacteria. These otherwise simple organisms use quorum sensing, in which individuals communicate information by chemical signals that pass between members of the same species, and occasionally between those of different species.

What the bacteria are able to read by chemical communication is the condition and density of the population to which they belong. With this information, the individual bacterium "decides" the rapidity of its movement, the rate of its reproduction, and, in the case of pathogenic species, even the virulence of its impact on the host in which it lives. In some incidents bacteria choose to form stable groups shielded by protective membranes and crusts, structures called biofilms.

Bacteria have thus been found to be social to a degree almost unimaginable to scientists a generation ago. But of course the microbes are also mindless. Whether a persistent group of any kind of organism can evolve further than the microbial depends on the complexity of the individuals that compose it. Consider a pod of bottlenose dolphins feeding on a school of anchovies. The small fish that serve as prey enjoy the same advantages of group membership as do starlings. Milling sleek and swift in masses of up to millions, they find food more quickly. Their combined mass also gives to each individual a higher average benefit of protection against the much smaller population of dolphins. Each school of anchovies is like a gigantic fish that its enemies are able only to nibble.

Dolphins feeding on anchovies push back at the mass schooling of their prey through a cooperation of their own. By swimming in groups around the anchovies in what appear to be quite intelligent coordinated movements, they corral the school into a tight sphere. There they are able to seize the fish singly or in small groups more precisely, like turning an apple about before taking healthy bites.

Social mammals such as dolphins and primates, and we ourselves, with larger brains, live in a social world more sophisticated than the most well-organized bacteria and schooling fish. They can think ahead, a process that automatically invites a higher level of order. They learn to recognize other group members personally. They can thereby plan their own actions with reference to both the group as a whole and individuals within it. What emerges in the mind of each animal is a spread of possible options, and from these an investment strategy comprising trade-offs in the exchange of personal information. Each group member learns when to cooperate or compete, and when to lead or follow.

Investment strategies, generated by natural selection at both the individual and group levels, can be viewed as the rules of games played as a series while each is instinctive in origin. (What is best for me? What is best for my group, thereby best for me?) It is acquired as genetically predisposed learning during interaction with other group members. In the Old World monkeys and apes with the most advanced and best studied societies, the rules for males are commonly as follows:

Young Males of Old World Monkeys and Apes: How to Succeed

✳ *If you are still too young and small to challenge higher-ranked mates, wait, plan, form coalitions with others of equal status.*

V *Cultivate favors from mentors of higher status.*

V *If you see a role in group activity poorly attended, such as forager or sentinel, fill it if possible, learn from the experience, and use it to lead young males of similar age and rank.*

* *Either dominate other males and copulate with females located near the center of the group, or else (usually) hide and attempt copulation with a single partner.*

Persistent, well-organized animal groups are potentially immortal. Dying members can be replaced indefinitely by newborn substitutes or individuals permitted to join from other groups. In one remarkable example, a mixed flock of roving, insect-eating birds censused in the rain forests of French Guiana was observed to persist across an interval of at least seventeen years. It comprised many bird generations, each of which remained faithful to roost sites, home range, and species composition.

Such elementary societies nevertheless remain mortal. They cannot anticipate every potentially deadly predator, every failure of their food supply. Vast numbers must have come and gone for the past half-billion years. Among them, an extremely few have evolved to the next and highest level of all. This is eusociality, in which the colony is divided into a "royal" caste specialized for reproduction, and a nonreproductive "worker" caste that performs the labor of the colony. Eusociality may be a relatively rare condition in evolution, but it has resulted in the most advanced levels of individual altruism and social complexity. It has conferred ecological dominance on the land by some of the species that possess it, particularly the ants, termites, and humans.

5

THE FINAL STEPS TO EUSOCIALITY

Eusociality did not evolve from species that might at first seem most likely to succeed. No matter how persistent and structured are the swarms, flocks, schools, pods, packs, herds, crashes, hordes, and murmurations, not one to the best of my knowledge has ever given rise to colonies divided into reproductive and nonreproductive castes. Biologists like myself have had to look elsewhere for clues to the origin of these most advanced of all societies. They have located the ancestors in species with an entirely different life cycle and mode of social behavior from others that look more promising yet turned out to be less successful.

Furthermore, despite expectation from the dramatic ecological success it confers, eusociality has arisen only rarely. The evidence shows that the process typically began when some members of a group, usually a family, practiced altruism to a degree already exceeding that occurring between ordinary parents and their offspring. It consisted of an early and abrupt withdrawal from personal reproduction on the part of at least a small number of individuals. The final step thereafter was not a consequence of the close kinship of family members, as many researchers have supposed. The reverse is true: close kinship within the group typically followed from the origin of eusociality. I'll now explain this reversal as conceived by me and a few others, starting with a background of the spectacular success of insects throughout the history of terrestrial life.

Paleontologists, studying fossils, together with sociobiologists, working on living species, have searched far and wide for evidences of eusociality. Their efforts have focused on insects, which consist of over a million known species. Within this great array about twenty thousand have been found to be eusocial. The latter comprise mostly ants, social bees, social wasps, and termites. But there also exist eusocial species of beetles, thrips, and aphids. The list may sound long, but it comprises only 2 percent of the contemporary million insects known to science.

By the 1970s, we had come to recognize that the origin of eusocial societies has not only been uncommon, it also has been a relatively recent event in the long evolutionary history of the insects and other kinds of animals.

The relative scarcity and geological youth of eusociality may be due to its being the last of the great evolutionary innovations that founded the modern insect world as a whole. The earliest was the origin of insects themselves. All arose and remained thereafter as land-based animals. If you would like to see primitive insects, turn over a few rocks in forest or meadowland and (perhaps with an entomologist at your side) look for springtails, proturans, silverfish, bristletails, all flightless insects similar to their ancestors.

The second innovation by insects as a whole was winged flight, making them the first of all animals to gain mastery of the air. Then came the ability to fold the wings over the back, allowing some species not only to fly with their wings outspread but also to scurry to safe cover when threatened by predators. If cockroaches come to your mind, yes, they were among the first insects to achieve this capability. The next innovation was complete metamorphosis, in which an immature form evolved that was radically different from the adult in anatomy and way of life. For example, a caterpillar, after consuming the leaves of a plant, metamorphoses into a butterfly that imbibes its nectar. Metamorphosis grants the same individual access to more than one food source and even more than one habitat. A metamorphosing dragonfly, for example, switches from aquatic swimmer to winged flyer.

Finally in the succession of major evolutionary advances came the eusocial colonies, which arose only after repeated major diversifications of the insects and other arthropods during the first 325 million years of their history. Until then no ants, termites, or their equivalents arose, so far as we know, to fill even part of the terrestrial world.

The earliest known fossil insect of any kind dates to the early Devonian, about 415 million years before the present. The land soon thereafter (in geological terms) filled with a growing roster of taxonomic insect orders. By the end of the Paleozoic Era, 252 million years before the present, insects as a whole had acquired a remarkably modern cast. Of the twenty-eight taxonomic orders alive today, fourteen were present then. As the Paleozoic Era (the age of coal forests and amphibians) ended and the Mesozoic Era (the Age of Reptiles) began, the survivors included many kinds of insects familiar today: barklice, dobsonflies and other neuropterans, stoneflies, beetles, and hemipterans such as treehoppers and shield bugs. These ancestors anatomically resembled their modern descendants, but they lived in a radically different world. If you could travel back to a coal swamp in the late Paleozoic, you would think odd the lycopsis trees shaped like royal palms, horsetails, and tree ferns. You would certainly be frightened (or at least should be) by the monstrous hungry, stubby-legged labyrinthodonts waddling toward you. But the insects buzzing around your head and crawling up your leg would, upon inspection, make you feel right at home.

Across the full history of Paleozoic evolution, spanning 415 million years to 252 million years before the present, the rich fossil record left no known evidence of eusocial life. Of course, that perception may change with further research—fossil records are always far from complete. Species living in eusocial colonies may have existed in sparse or local populations still missed by the fossil hunters. Others could have evolved in hidden niches, as do contemporary eusocial bark beetles and gall-forming thrips. Still, no trace of any kind has yet been detected anywhere in the rich Paleozoic fossil

A tableau of several animal lines among the eighteen known to have attained eusociality. African mole rats (center) are surrounded by (top, then clockwise) a social wasp, apid bees, termites (huge queen attended by workers), ants, and a bumble bee.

deposits of an anatomically distinct worker caste, the hallmark of eusociality.

This evidence, even though negative, deserves attention because of its relevance to our general understanding of advanced social evolution. It raises the important questions of *why* eusociality has been rare and *why* also it came so late in geological time.

A sparse diversity of eusociality in the contemporary insect world is further evidence of its rarity in geological history. Only seventeen independent origins among all animals are known that created the eusocial colonies in existence today. Three of the independent lines are alphaeid shrimp, found in shallow marine waters of the tropics (and the only marine eusocial animals known of any kind). The alphaeid queens and workers create nests by excavating burrows in living sponges. Two additional independent lines gave rise to eusociality in the known vespid wasps, familiar examples of which are the hornets, yellow jackets, and paper wasps. Two more eusocial lines have been discovered among the bark beetles, members of the taxonomic family Scolytidae. (More technically, the scolytids are nowadays placed within the weevil family Curculionidae.) Scolytids comprise a huge assemblage of species best known from the few that have become scourges of coniferous forests. An additional two eusocial species are the naked mole rats of Africa, blind, hairless vegetarians that live in deep burrows in the soil.

There are seven remaining lines of this advanced form of society that originated separately. One line each attained their modern form in ants, termites, sphecid wasps, allodapine bees, augochlorine bees, thrips, and aphids. (A cockroach species from the Mesozoic Era, given the technical name *Sociala perlucida*, has been interpreted as the caste of a eusocial species, but the claim is far from proven.)

Finally, a plausible case can be made for eusociality in human beings. The strongest evidence is the postmenopausal "caste" of grandmother helpers. In addition there is the readiness with which individuals join professions and callings useful to society but counter to their own reproduction. Given that homosexuality is uniquely valuable to so many societies, it is not unreasonable to view homosexuals as a eusocial caste, and in the highest possible sense. In further witness is the prevalence of monastic orders in organized religions around the world. In yet another venue must be included the formally established and respected *berdache* system of the early Plains Indians, in which males dressed and performed as females. It should be kept in mind that the propensity toward homosexuality has a partly genetic basis, and further appears to benefit relatives and larger groups, making its genes more likely to survive. The evidence is indirect but strong: the frequency of homosexual-propensity genes in human populations is above the level expected from mutation alone, a sign that the propensity has been favored by natural selection. The level, in other words, is too high to be explained solely by random changes in genes that affect sexual behavior.

Additional evolutionary lines of eusociality will almost certainly be found. They are most likely to exist somewhere among the myriad living species of insects and other arthropods. But I doubt that the number will ever rise to more than a tiny fraction of all the animal evolutionary lines and all of the species within them. The overwhelming fact, to repeat and keep in mind, is that the known species of ants, termites, and eusocial bees and wasps together, having attained world dominance in numbers, biomass, and ecological impact, still comprise a tiny fraction of the one million known insect species. Further examples of eusocial species can be expected to be not only rare but also marginalized into small, specialized niches.

The timing of the insect conquest is crucial. The origins of the living insect eusocial lines alive today were scattered across the Mesozoic and Cenozoic Eras. Termites were first among them, projected to have evolved from cockroach-like ancestors during the Middle Triassic to Early Jurassic (237–174 million years before the present). The eusocial corbiculate bees, particularly the bumble bees (tribe Bombini), honey bees (Apini), and stingless bees (Meliponini), evidently originated variously toward the end of the Cretaceous Period, as far back as 87 million years ago. The origin of eusociality in halictid bees occurred during the mid-Paleogene Epoch, around 35 million years ago. The ants appeared, evidently from a single aculeate wasp ancestor, during the Cretaceous Period, about 140 million years ago.

By the Paleogene Epoch and likely further back during the very Late Cretaceous, most or all of the recognized contemporary twenty-one ant subfamilies had also separated.

Why was eusociality so late in coming? And *why* has it remained so infrequent, especially when as a whole it has proven so ecologically successful? Numerous candidate evolutionary lines and environmental opportunities to advance into eusociality have existed on the land as well as in the fresh and shallow marine waters as far back as the first terrestrial invasion by multicellular life. At least tens of thousands, more likely hundreds of thousands, of insect species were present and diversifying during the late Paleozoic and early Mesozoic Eras. During that time they occupied a wide range of environmental niches. The Pennsylvanian tree fern *Psaronius*, for example, was host to at least seven insect groups with different feeding habits, including variously external foliage consumption, piercing and sucking, stem boring, galling, spore consumption, and ingesting litter and peat at the base of trees. Many types of life cycles and dispersal mechanisms have persisted and originated from that time onward. Also, various degrees of relatedness, from clonal to unrelated, probably were present in groups of individuals, just as they are today in modern lineages of Paleozoic origin.

At the present time social aggregations of ancient provenance, still short of eusociality, occur in different patterns and degrees of complexity in a majority of the insect orders. Massed offspring are tended by mothers and sometimes fathers as well. In a few cases these offspring are led by their parents from one place to another. According to species, the young are either protected by nests or kept in the open. In particular, long-term care and protection of young have been observed in membracid treehoppers, scutellerid jewel bugs, belostomatid giant water bugs, gall-dwelling aphids, tingid lacebugs, praying mantises, earwigs, and argid sawflies. Tight masses of larvae, adults, or both, in some cases capable of organized movement, occur in species as diverse as gyrinid whirligig beetles, psocopteran barklice, embiidine webspinners, noctuid and lasiocampid moths, lubber grasshoppers, cockroaches, and tenthridinid and pamphilid sawflies.

From this huge array of subsocial insects and other animal species has arisen the very small set of independent lines of living eusocial species. The origin of advanced societies is obviously not correlated with the degree of kinship within families and other tightly knit groups. The key to their origin is very different. It is that all of these lines, with no known exception, first attained the relatively rare preadaptation of progressive care of the young in nests from egg to maturity by regular feeding or inspection or both, combined with persistent protection against enemies.

The overall pattern in the emergence of eusociality began to be revealed over half a century ago by the pioneering studies of Charles D. Michener of the University of Kansas on bees and of Howard E. Evans at Harvard University on wasps. Both were my mentors and greatly affected my early studies on ants. The full sequence, worked out from their research and continued to the present time by many experts, is the following. In the beginning, adults of many species build nests, store cells with pollen or paralyzed prey, lay eggs, seal the nest, and depart. In a smaller number of spe-

The evolution of social behavior that led to eusociality in wasps. (Top, then clockwise) In an example of the first step, a bethylid female wasp stings and paralyzes a tiger beetle larva, lays an egg on it, then leaves the larva in place for her offspring to consume. In the second step, a sphecid wasp stings and paralyzes a black widow spider to take to her own nest as food. A vespid wasp paralyzes and brings a succession of prey to feed her larvae. Finally, the mother and daughters stay together as a eusocial colony, she as queen and they as workers.

cies, beginning the second phase, the adults build the nest and lay eggs, then care for the young throughout their development by periodic feeding or by nest cleaning, or by both. Finally, in a much smaller number still, now classified as primitively eusocial species, the mother and adult offspring stay together at the nest, with the mother continuing to lay eggs as the primary reproductrix, and the offspring foraging and laboring as nonreproductive workers.

The sequence inferred by analysts to have produced the advanced societies of ants and the eusocial wasps is the following, also shown in the accompanying figure. In early Cretaceous times, at roughly 200 million to 150 million years ago, aculeate (stinging) wasps preyed on insects living in soil and leaf litter. If they were like modern species of wasp families such as the Bethylidae, Mutillidae, Pompilidae, Sphecidae, and Tiphiidae, which are familiar companions on a summer nature walk today, many would have specialized to feed on spiders and the larvae of beetles. After mating, the female located prey by odor, attacked and stung them with a paralyzing venom, laid an egg on each, and left them for the hatching larvae to consume. Modern bethylid wasps of the genus *Methocha*, for example, invade the burrows of tiger beetle larvae, sting the inhabitant, lay an egg, and leave the victim and egg in place.

A smaller array of aculeate wasps, derived from these more primitive huntresses, carry the paralyzed prey to a nest they themselves have prepared, lay an egg, seal the nest shut, and depart to repeat the procedure elsewhere. Among the most familiar examples are the various species of mud dauber wasps (Sphecidae) that build earthen nests under bridges and in the eaves of houses.

A still smaller group of aculeate wasp species remain with the brood, bringing fresh prey progressively as their larval offspring grow to maturity. When the young reach maturity, they and their mother disperse singly.

Finally, in a very small group of progressively provisioning species, including the ancestors of the ants and eusocial wasps, the mother and her offspring remain together, forming a eusocial colony.

This sequence of steadily diminishing sets of evolutionary lines and species points to an uncommon adaptation in the life cycle, as opposed to close genetic relatedness among the members of the founding group, which is often speculated to be the ruling precondition for the origin of eusociality. In fact, it follows (as I have stressed) that close kinship is not the cause of the origin of eusociality, but its consequence. All that is needed to cross the threshold from the solitary way of life to the eusocial way of life is the silencing by mutation of one or more alleles that prescribe the tendency of parents to care for their offspring at first, then separate and disperse when their offspring reach maturity.

A second documented preadaptation that favors the theory of this transition to eusociality is the propensity of solitary bees to behave like eusocial bees when experimentally forced together by scientists. The coerced partners proceeded variously to divide labor by taking up foraging, tunneling, or

Eusociality, the organization of a group into reproductive and nonreproductive castes, occurred in only a tiny percentage of evolving lines, then relatively late in geological time, and almost entirely on the land. Yet these few, leading to the ants, termites, and humans, have come to dominate the terrestrial animal world.

guarding of the nest. Furthermore, the females engaged in leadership, one bee leading and the other following, a behavior seen in eusocial bees. This elementary division of labor appears to be the result of a preexisting behavioral ground plan, in which solitary individuals tend to move from one job to another after the first is completed, a straightforward way to raise their personal offspring. In eusocial species, the algorithm is transferred to the avoidance of a job already being filled by another worker. It is evident that progressively provisioning bees and wasps are "spring-loaded" (strongly predisposed, with a specialized trigger stimulus) for a rapid shift to eusociality, once group selection (group competing with solitaires and other groups) favors the change.

This line of reasoning, how and why advanced social behavior originated, is typical of the way scientific theory is created in general. A successful theory fits independent tested facts like pieces of a jigsaw puzzle. Here the results of the experiments on solitary bees forced together fit the fixed-threshold model of the origin of labor division proposed by developmental biologists for the emergence of the phenomenon in established insect societies. The theory in this case posits that variation, sometimes genetic in origin and sometimes a result of learning, exists in the response thresholds associated with various tasks. When two or more individuals interact, it says, those with the lowest threshold are the first to begin the task. The activity then inhibits their partners from doing unnecessary work, by causing them—instinctively—to move on to whatever other tasks are available. Thus, once again, the impact of a single flexible gene change, inhibiting dispersal of group members from the natal nest, would seem to be enough to carry preadapted species across the threshold to an advanced instinctive social order.

Comparative studies in the field and laboratory have revealed that from the moment of the evolutionary origin of animal eusociality, the worker is in a tug-of-war between its own interest and that of the colony to which it belongs. As colony-level organization becomes more important to the success of the alleles prescribing the organization, individual worker survival and reproduction become increasingly less important. Finally, in obligatory eusociality, the capacity for worker reproduction within the genome ceases, creating the ultimate superorganism. Extreme superorganisms in the insect world, in which the female workers lack any capacity to reproduce, are found for example in many kinds of ants, including doryline army ants, *Atta* fungus growers, and five other major groups, the genera *Solenopsis*, *Pheidole*, *Monomorium*, *Tetramorium*, and *Linepithema*. In these species workers lack ovaries altogether. On the other hand, the capacity of workers to reproduce has returned, or at least been augmented, in a few clades of species by secondary evolution, allowing individual workers to assume the role of queen. At the extreme superorganismic phase, the level of selection becomes the genome of the queen, and the workers are more precisely viewed as the robotic extensions of her phenotype.

6

GROUP SELECTION

BIOLOGISTS HAVE SCANNED A HALF-BILLION years of land-based evolution for evidence of advanced animal societies. From this knowledge they have tried to acquire a better understanding of our own species. But they have been stymied by a genetic mystery of the first magnitude.

The mystery consists of two parts. The first, which I've addressed in the previous pages, was perceived and largely solved, at least in very general terms, by Charles Darwin in *On the Origin of Species* (1859) and *The Descent of Man* (1871). It is, how can advanced societies evolve, when many individuals serving the society cease to reproduce? Put in familiar terms, how did altruism come into existence? The solution Darwin suggested is what we today have refined into the theory of group selection. Some members of the group, the theory states, can shorten their lives, or reduce their personal reproduction, or both, if the advantage their sacrifice provides the group gives sufficient advantage over other, competing groups. The altruism gene then spreads through the population of the group by mutation and selection. It is hastened but not driven by close kinship of the group members. Close kinship often follows but does not precede the spread of the altruism. The models of population genetics show that even the average presence of a single hereditary altruist in a group, whether or not the group members are kin, results in an increase in the population of such groups as a whole.

This perception brings us to the second puzzle. Why has the origin of eusociality, featuring division of labor based on altruism, been so rare in evolution? The answer must lie somewhere in the following precondition for it to occur: a mother or small group progressively raising young within a fortified nest. This requirement is actually very common in nature, yet in the vast majority of cases it has not yielded eusociality. So the more relevant question is, What prevents the final step? The identity of this inhibiting agency could solve the second chapter in the mystery of eusociality.

I believe the answer lies in the great difficulty biologically inherent in the final step, as follows. Consider a small colony consisting of a mother (perhaps with a father helping her) plus her brood of freshly emerged adult offspring. An ordinary life cycle ends at this point. A new life cycle begins as the mother and her female offspring each disperse in independent solitude. The mother would either die or perhaps start a new brood, while each of the offspring mates and sets out to build a nest and become a mother herself.

Now suppose a knockout mutation, as little as one change in a single gene in our hypothetical scenario, occurred that cancels the dispersal of the little family. (Knockout mutations, which negate other mutations, are relatively common and moreover have been extensively used in genetic research.) We know that if a group of mature females is kept together experimentally, the one first on the scene and already inseminated, in other words the mother, will dominate the group and serve as the egg-layer, while the others will serve as workers.

Thus, once the preliminary adaptation, comprising the construction of a fortified nest plus progressive care of the young, is in place, it would in principal be elementary to evolve the one step further to eusociality. But although this advance outwardly seems easy, it is rarely taken in nature. Why? The explanation presenting itself is that although a mutation in one gene or a small ensemble of genes can make a eusocial colony, the whole remainder of the original genome remains adapted to solitary life. The daughters may be newly created workers in their instinct to stay home, but they are programmed in all other respects to live as solitary organisms. They are not prepared to communicate with one another or divide labor among themselves in nest building, nursing, and foraging. Thus encumbered, the unchanged group cannot effectively compete with either their solitary peers or the colonies of other, successfully evolved eusocial species.

There is now abundant documentation of the fundamental genetic changes that underlie the evolution of eusociality. In 2015 an international team of fifty-two researchers led by Karen M. Kapheim and Gene M. Robinson at the University of Illinois reported on the genomes of ten species of bees representing multiple independent lineages in different stages of evolution. The social sequence collectively begins in solitary life and ends in complex eusociality. Each of the lineages represented was found to possess its own pathway of genetic evolution, but all those achieving sociality displayed the same basic pattern of change. They all displayed an apparent increase in the amount of neutral evolution as a consequence of relaxed natural selection, rising with social complexity, and a concomitant decrease in the diversity and abundance of transposable elements. Overall, to put this admittedly technical matter as simply as possible, advanced social organization entails an increase in the complexity of the gene networks affecting social behavior. Advanced social behavior does entail a basic change in the genetic code.

During the 1950s, the British entomologist Michael V. Brian and I independently provided evidence of the intricate mechanisms in larval development that create the worker and reproductive castes—hence eusociality—in ants. In the European species, *Myrmica ruginodis*, Brian found that each larva has the potential to mature either as a queen, possessing a large body, wings, and fully developed ovaries, or as a worker, which is smaller, wingless, and sterile. There exists a threshold size—a "decision point" at which the larva is destined to complete its growth and metamorphose into either an adult queen or an adult worker. Brian found that the fate of the growing *Myrmica* larva, whether to a queen or worker, depends upon a combination of factors, namely the size of the egg from which the larva hatched, the size it reaches by a certain point in its growth, the presence or absence of the colony's mother queen, the age of the mother queen, and finally whether it lived as a young larva in winter and had been chilled prior to rapid growth in the spring. All of these factors together provide the colony with a supply of virgin queens to be released in the nuptial flights during later warm weather. Each has the potential to mate and start a new colony of her own.

Much later, in 2002, Ehab Abouheif and his fellow researchers at Montreal's McGill University, working at the basic level of the genome, discovered that the capacity of ants generally to produce winged queens is dependent on modified genes possessed by females. The genetic network affecting development to the adult stage is conserved in the winged queen caste but disrupted in the wingless worker caste. The worker, in short, loses its potential genetic endowment.

A lot of information now fell into place. In 1953, I had measured species from all of the forty-nine ant genera in the world known to possess more than one subcaste of worker, that is, a workforce divided into minor workers and major workers, the latter sometimes referred to as soldiers. Many of these species have intermediates (media workers), and a few have an even larger third caste called supermajors. During the origin of advanced social organization, the added subcastes require not only one or two additional decision points in development of the larvae, but a regulation of their relative members in different stages of colony growth.

The regulation is the equivalent of the division of labor in humans based on different occupations plus cultural regulations on the number trained in each occupation.

Thus have emerged the empires of ants and men.

The only way to acquire the necessary genetic changes and overcome the solitary-genome barrier is by group selection, which has the power to generate gene-based altruism, division of labor, and cooperation among members of a group. This higher level of natural selection is already a well-documented and directly observed force in ants, and in social insects generally, not just during colony founding but also during competition among mature colonies. Conflict may arise by direct physical interference, resulting either in retreat or complete destruction ("myrmicide" for the ants, to coin a term) of the losing colony. Competition between colonies does not, however, consist exclusively of combat and predation among colonies. It also includes preemption of new foraging sites and chasing off or killing rivals, as well as superior ability in the harvesting of nest materials and food. Theoretical and experimental studies have demonstrated that all of these heritable colony-level endeavors are dependent primarily on the rate of colony growth and mature colony size, both of which use genetically determined group-level phenotypes. The number of participating workers alone, all else being equal, has a profound effect on the colony's metabolic growth rate. With more workers colonies grow faster, produce more queens and males, and reach larger mature size. The relationships mirror the metabolic scaling laws for mass and physiology of individual organisms. Mathematical modeling demonstrates that the most critical demographic factor in the competitive growth of insect colonies is likely to be the initial fecundity of the founding queen.

At this point it's important to review the process of group selection as defined within the tested principles of population genetics, and through them correctly explain social evolution. It deserves emphasis. *For group-level traits as for individual-level traits, the unit of selection is the gene that prescribes the trait. The targets of natural selection, which determine whether genes do either well or poorly, are the traits prescribed by the genes.* An individual in a group that competes with other members for food, mates, and status is engaged in natural selection at the individual level. Individuals that interact with other group members in ways that create superior organization through hierarchies, leadership, and cooperation, are engaged in natural selection at the group level. The greater the price exacted by altruism and the resulting loss to the individual's survival and reproduction, the larger must be the benefit to the group as a whole. The evolutionary biologist David Sloan Wilson (no relation) has nicely expressed the rule for the two levels of selection as follows: within groups, selfish individuals win against altruists, but groups of altruists beat groups of selfish individuals.

The actual process of group selection has in recent years been illuminated by the study of examples of the process

Group selection is natural selection of alleles (alternative forms of the same gene) that prescribe social traits. The traits favored by natural selection are those that entail the interaction of individuals within groups, including the initial formation of the groups. As groups of the same species then compete, the genes of their members are tested, driving social evolution by natural selection up or down. A rich documentation of this process has been provided by both natural history and experimental studies.

under natural conditions. An appropriate example to begin with are the wolves of Yellowstone National Park, which have taught us so much about ecology and sociobiology. Recent studies by Kira Cassidy of the University of Minnesota and her coworkers have revealed that when groups came into territorial conflict, larger packs, averaging at the time of the study 9.4 wolves, won over smaller packs, averaging 5.8 members. Furthermore, those with a higher proportion of adult males were more likely to defeat those with a lower proportion. And finally, the presence of a male or female six years or older (the average lifespan in Yellowstone is four years) tipped the balance still more.

To witness group selection progress in the greatest variety of venues, let us next turn to the invertebrates. A particularly striking example exists in the queen cooperation and rivalry of the imported fire ant (*Solenopsis invicta*) evaluated in detail by Walter R. Tschinkel in his classic treatment, *The Fire Ants* (2006). Following their nuptial flights and aerial insemination, single queens often gather in groups of up to ten or more, build a little nest together, and then rear the first brood cooperatively. This unusual behavior is manifestly driven by group selection. In a world of intense and deadly competition, fewer than one in a thousand queens live to be mother of a colony populous enough to produce daughter queens. Field studies have shown that the size of each colony is supremely important for its survival, and this is conspicuously the case for very young colonies. In the laboratory, groups of cooperating queens rear more workers per queen and at a faster rate than do solitary queens.

When the fire ant workers reach maturity, they begin to eliminate the queens one by one, spread-eagling them and stinging them to death, until only one is left. They do not spare their own mother. The winner, identifiable by her pheromones, is the most fecund, hence the one most able to foster rapid growth of the colony as a whole. The workers cannot afford the cost of supporting losers, even if their own mothers must die. Group selection in this instance prevails decisively over individual selection.

The great diversity of ants worldwide, comprising over 15,000 known species, makes them ideal subjects for identifying the factors of social evolution in a comparative manner. The central questions their study addresses can be boiled down to three in number. First, who or what controls the number of workers in the colony, second how is that separately accomplished, and third what forces of natural selection are responsible.

Rapid DNA mapping has made analysis of the social factors of ants more accessible by experiments using entire colonies. It has added to the picture of group selection as the grandmaster of social evolution in these insects. In the phenomenon of policing, worker ants harass and sometimes go so far as to execute nestmates who lay eggs in competition with the mother queen. In the past, policing was usually explained by inclusive fitness theory based on degrees of kinship among the workers. It was widely believed, as a principle, that harassment is strongest against would-be usurpers who are most distantly related to the punishers. However, the same effect can be explained by individual differences in closeness to the overall colony odor. As Serafino Teseo, Daniel Kronauer, and their fellow researchers at Rockefeller University recently demonstrated, the increase of colony efficiency can explain policing in full. They found that colonies of the tropical ant *Cerapachys biroi*, which are clonal, hence with genetically identical workers, nevertheless still practice policing. The explanation of this phenomenon lies in another domain of biology, as follows. Growth and regulation are regulated in colony cycles induced by the larvae. During part of the cycle, the ovaries of the adults are shut down in response to cues from these immature, grublike members. Individuals who fail to respond to the cues, and thus disrupt the cycle, are harassed and sometimes executed. In a series of ingenious laboratory experiments, the researchers assembled two kinds of queenless colonies of *Cerapachys*, one a typical clone and the other a chimera (different parents) of two genetically different colonies put together in the laboratory. The single-clone colony outcompeted the chimera, evidently because the chimera colony produced many individuals that chose not to work but to reproduce. Their efforts overrode the normal reproductive cycle and thereby reduced efficiency at the colony level.

A parallel, independent study by Shigeto Dobata and Kazuki Tsuji of the University of the Ryukyus used a second clonal ant species, *Pristomyrmex punctatus*, to reach a similar result. Because there is no queen, all the workers participate in laying eggs and rearing the young to maturity. As in the queenless *Cerapachys* colonies, no advantage accrues to the individual by laying her eggs. All of the immature members are genetically identical, and all are reared as a single egalitarian community. Each is a potential mother as well as an exact copy of all the other mothers. In the field, colonies are infiltrated by genetically different workers from other colonies. These aliens cheat by laying more eggs than the native residents and by avoiding labor. In the laboratory, cheaters overall contributed more offspring per capita. When groups composed entirely of cheaters were put together, they failed to produce any offspring at all.

What are we to make of this odd phenomenon? Kinship matters to this extent in the Pristomyrmex ants: the working mothers of clonal colonies recognize members of other clonal colonies as aliens. When cheaters invade nests of different colonies, they act as social parasites, invading and exploiting the labor of another species. An equivalent among birds are the cuckoos, which sneak their eggs into the nests of other species.

In 2001, Patrick Abbot and coworkers at the University of Arizona were the first to report a similar phenomenon in eusocial aphids. The species studied form highly organized colonies, going so far as to produce a soldier caste. They are also clonal, hence not subject to a social order shaped by kinship. In at least one of the species (*Pemphigus obesinymphae*), the colonies are not always purely colonial but instead often vitiated by invasions from other clones. The intruders then act as parasites. They do not take upon themselves the risky defense of the host colony. Instead, they alter their physiology selfishly to become reproductives.

During the history of sociobiological research, which combines natural history and genetics, such surprising new patterns in the life cycles of social species are coming ever more frequently to light. One of the most remarkable and instructive is the reproductive queue worked out in social wasps by Raghavendra Gadagkar and his collaborators at the Indian Institute of Science in Bangalore. Colonies of the Asian wasp *Ropalidia marginata*, the researchers found, seem outwardly simple in social organization but are actually guided by sophisticated rules of cooperation. The workers of a *Ropalidia* colony are physiologically capable of reproduction, but all defer to the reigning queen. The ruler in this case is not the most aggressive individual, nor is she the head of a dominance hierarchy. Nevertheless, she enjoys a complete monopoly on egg-laying. The *Ropalidia* can be said to be a benevolent autocracy. When the queen is removed, one of the workers becomes temporarily hyperaggressive toward her nestmates. Her displays and threats are almost never challenged. Once established, the new queen returns to her original docility. Her ovaries develop, and she begins to lay eggs. She is thereafter the exclusive reproductrix. When she in turn dies or is removed by the researchers, she is soon replaced by yet another worker ruler, who has somehow already been primed for the job. When that successor departs, another takes her place, and so on ad seriatim. The colony moves more or less peacefully through a queue of cryptic (to humans) heir apparents.

It turns out that the new *Ropalidia* queen in each succession is not the most closely related to the other workers. Rather, she is typically the oldest. The entire procedure appears mediated by peacemaker pheromones. The royal succession is thus manifested as a colony-level adaptation. It eliminates almost all violent and destructive conflict. It also reduces the risk of both internal anarchy and invasion by usurpers from other colonies. *Ropalidia marginata* colonies are thereby rendered immortal in theory, even though—given the vicissitudes of the environment—they are almost always short-lived in practice.

Another, very different kind of selection presumably acting peacefully at the group level has been documented in a separate line of primitively eusocial social wasps. Among all nineteen such species observed in nature, lone females were found during one study to be at high risk both on the nest and while foraging. From 38 to 100 percent of the foundresses monitored in different samples failed before the first brood emerged. In a separate study it was found that when queens among multiple foundresses of individual colonies in at least two of the wasp species, *Liostenogaster fralineata* and *Eustenogaster fraterna*, disappear, the orphaned helpers rear her brood on to maturity—whether they are kin or not. At the same time the helpers lay eggs to start their own brood. Thus they create an "insurance-based" advantage to all of the cooperators through the perpetuation of eusociality.

As the exploration of animal societies has deepened, sociobiologists have encountered ever more pathways of evolution, some so surprising as to border on the bizarre. At least a few of these aberrations occur in spiders. Investigators studying eusociality and its precedents have hoped someday to find examples of eusocial spiders. Social spiders sharing the same large webs are well known in at least two independent phylogenetic lines, but none of their species has been found to produce reproductive and worker castes.

However, the spider-nest inhabitants do show "personality" differences evidently sustained by group selection. The phenomenon occurs in *Anelosimus*, a genus of spiders with a worldwide distribution and rich diversity of local species. They are members of the cobweb spider family Theridiidae, which includes the black widow. Many, like their notorious cousin, wear bright patterns of color on their abdomen. What are more notable, however, are those that form colonies, creating an arachnophobe's nightmare of up to thousands of hungry cooperative females suspended together in communal webs. Jonathan N. Pruitt and his collaborators at the University of Pittsburgh found that in colonies of the New World species *Anelosimus studiosus*, females are composed of two principal "personality" castes. The first aggressively participate in the capture of prey, the building of the web, and defense of the colony. The second are relatively docile and concerned with parental care, including protection of the large spherical egg masses. The aggressives are more effective at capturing food and repelling invaders, while dociles were better at rearing large quantities of brood. The difference in personality appears to have at least a partly genetic basis, but the two types live together in relative harmony.

The advantage of *Anelosimus studiosus* colonies for science is that they can be experimentally constructed with different ratios of the caste from selected sites in nature, and placed in other sites with different environments, in order to see how the artificial colonies would adapt. In doing so, Pruitt and his fellow researchers were in effect testing for the occurrence of group selection. The results were positive: in both the original and new sites, each of the colonies shifted over two generations to the aggressive–docile ratios found at its original site.

Finally, it is with the termites and their apparently immediate ancestors that we are allowed to witness almost directly the breakout by group selection from the eusociality threshold barrier—and thence into full eusociality.

It is generally agreed among experts that termites descended from cockroaches. Evolutionary biologists temper such assessments by saying instead that two closely related kinds of insects came from a common ancestor. But the phylogenetic sequence in this case is so tight I believe it correct to say that termites are social cockroaches.

The living cockroaches closest to termites are large, wood-eating species of the genus *Cryptocercus*, found

*A colony of social spiders (*Anelosimus*) have captured a large beetle, and will share the food it provides. Also shown are the two "personality" types: hunters far in the figure and nurses protecting round egg masses close in the figure.*

in North America, eastern Russia, and western China. They superficially resemble hissing cockroaches (genus *Gromphadorhina*) of Madagascar, commonly used in laboratory research, as well as frightening "bugs" in Hollywood horror movies.

Cryptocercus are large for cockroaches. They survive not by sprinting away from enemies in the expected cockroach-in-the-kitchen manner, but passively, relying on the protection of their heavy chitinous armor. They carry thick exoskeletons and shieldlike forebodies, and move at a dignified pace on feet armed with spines. They defend permanently fixed homes in the decaying wood of dead trees and tree limbs. Christine Nalepa, at North Carolina State University, has recently drawn together the anatomical and genetic evidence of the closeness of the *Cryptocercus* to termites in their way of life and social behavior. Like the modern termites, she points out, they depend on specialized bacteria or other microorganisms that live in their guts. These symbionts digest rotting-wood cellulose and share the components with their insect hosts. Further, both the *Cryptocercus* cockroaches and termites rear their helpless young in part by feeding them the digested wood components through their anuses.

The *Cryptocercus* colonies, like termite societies, are in fact bound inextricably by the necessity of passing symbiotic wood-digesting bacteria or other microorganisms from one generation to the next. The *Cryptocercus* societies are typical families, with parents attending their offspring who grow to mature full size and become parents in turn. Termites, among the ruling insects of the world, also have families, but of a very different kind. Most of their offspring do not become parents. Instead, they develop as workers that support their parents and sibling workers. In other words, they generate a growing community. Thus has risen the condition of eusociality, the potentially most complex level of social organization, in which individuals are bound together by necessity to form a single reproducing unit. As eusocial cockroaches, termite colonies have shifted from the *Cryptocercus* stage of social life, forged primarily by individual-level selection, to the next level up, creating complex communities forged primarily by group selection.

Which brings us to a major controversy that has festered within sociobiology. It began with a thought experiment by the British biologist J. B. S. Haldane conducted and published in the 1950s.

This great scientist, in positing what later came to be called kin selection, illustrated the idea with the following thought experiment. Suppose you see a man drowning. If you try to save him, you have a 10 percent chance of accidentally drowning yourself. The genes that prescribe your social response are, let us assume, fully in charge. If the drowning man is a stranger, saving him is not worth the 10 percent risk of your own death and thereby all your personal genes. Your genes will not benefit if you were to take such a risk even if successful. However, if the drowning man is your brother, carrying half your genes, the 10 percent chance of all your genes would make the rescue worth the risk. That is, from the point of view of the genes, chancing a rescue is all that matters in the course of evolution by natural selection.

In composing this scenario, Haldane recognized that kin selection has the power to evolve altruistic behavior, hence eusocial societies like those of ants and people, and it depends on the closeness of the kinship of the altruist and the benefitted. The closer their kinship, the more genes they share, and hence the more of their genes are passed on to the next generation. An apocryphal quote about Haldane states, "He would lay down his life for eight cousins or two brothers."

In 1964 British geneticist William D. Hamilton suggested that kin selection could be a key factor in the origin of eusocial societies. He offered a formula of kin selection to show that it can favor a trait even if the trait is disfavored by ordinary individual selection, provided the benefit to others in a group, symbolized by B, when multiplied by degree of relatedness R, exceeds the cost to self. "Hamilton's rule," written as $BR - C > 0$, states the threshold above which true altruism can evolve.

The apparent achievement of expressing a complex process of social evolution in a physics-like formula has (until recently, at least) brought unusual popular attention to "Hamilton's rule—general" (HRG), and it is still often taught in elementary classes on sociobiology and evolutionary theory. Unfortunately, the passing of time has brought to light fatal weaknesses in the theory. Mathematicians and mathematically trained evolutionary biologists have been increasingly firm in rejecting it altogether as a correct or even useful scientific statement. For example, in 2013, as Martin A. Nowak, Alex McAvoy, Benjamin Allen, and I observed in an article in the *Proceedings of the National Academy of Sciences, USA*:

> *The mathematical investigation of HRG reveals three astonishing facts. First, HRG is logically incapable of making any prediction about any situation because the benefit,* B, *and the cost,* C, *cannot be known in advance. They depend on the data that are to be predicted. At the outset of an experiment,* B *and* C *are unknown, and so there is no way to say what Hamilton's rule would predict. Once the experiment is done, HRG will produce* B *and* C *values in retrospect such that* BR − C *is positive if the trait in question has increased and negative if it has decreased. But these "predictions" are merely rearrangements of the data that have been collected and already contain information about whether or not the trait has increased. In particular, the parameters* B *and* C *depend on the change in average trait value.*
>
> *The second astonishing fact of HRG is that the prediction, which exists only in retrospect, is not based on relatedness or any other aspect of population structure. A common interpretation of the terms in Hamilton's rule is that* R *quantifies the population structure, whereas* B *and* C *characterize the nature*

of the trait. But the derivation shows that this interpretation is wrong. All three terms, B, R, *and* C, *are functions of population structure, whereas the overall value of* BR − C *is functionally independent of population structure. Any information about who interacts with whom cancels out when calculating the value of* BR − C.

The third fact of HRG is that no conceivable experiment exists that could test (or invalidate) this rule. All input data, whether they come from biology or not, are formally in agreement with HRG. This agreement is not a consequence of natural selection, but a statement about a relationship between slopes in multivariate linear regression. This relationship between slopes has been known in statistics at least since 1897.

The same vacuity extends a fortiori to the abstract concept proposed by Hamilton called "inclusive fitness." Hamilton's rule is extended pairwise, individual to individual, to all of the members of a colony to determine how much the group as a whole is benefitted by the sum of all interactions. There is a small school of dedicated "IF theorists" who defend the idea, but no inclusive fitnesses have been measured in the real world, and it fails even in imaginative scenarios to make one possible.

I grant that I and other critics of inclusive fitness theory and its applications could be proved wrong, and measurements will someday be made or at least approximated indirectly. In that case Hamilton's extension of kin selection will indeed prove a major contribution to sociobiology. But for the present, advances in the understanding of the genesis of societies will have to be made the old-fashioned (and by far most interesting) way, by exploration in the field and laboratory, leading to generalizations hard-wrung from the databases.

7

THE HUMAN STORY

FOR NEARLY FOUR HUNDRED MILLION YEARS VAST numbers of large animal species (roughly defined as ten kilograms or more) evolved on the land, only to suffer extinction and replacement by their descendants. How many species arose, and how many vanished? Allow me to make a semi-educated guess. If as evidenced by the fossil record the average life span of a species together with the life span of its daughter species is on the order of a million years, and if conservatively a thousand such larger species are alive at the same time, then (perhaps!) a total of half a billion such species have lived on Earth throughout its history.

Only one of the multitude has reached the human level of intelligence and social organization. With this singular event, everything on the planet changed. Thereafter, there would be no other candidate and no further contest. The winner was one extraordinarily lucky species of Old World primate. The place was eastern and southern Africa. The habitat was a broad swath of tropical savanna, grassland, and semi-desert. The time was from three hundred thousand to two hundred thousand years ago.

The key events presaging the origin of humanity began

African primates observe a hunting pack of their world-changing human rivals cross the savanna.

five million to six million years before the present, when a single species of ape split into two species, thence two lines of multiplying species, one leading to contemporary *Homo sapiens*, the other to the two living chimpanzee species, the common chimpanzee (*Pan troglodytes*) and its smaller, more human-like cousin, the bonobo (*Pan paniscus*).

Both evolving lineages descended to live primarily, but not entirely, on the ground, the prehuman species more so than the pre-chimpanzees. The pre-chimpanzee species could run awkwardly on its hind legs alone or on all fours, hands folded and knuckles dragging. At no later than 4.4 million years before the present, the oldest known human antecedent, *Ardipithecus ramidus*, walked on elongated hind legs while retaining long arms and the ability to climb and move about in trees.

With this first step, so to speak, toward a terrestrial existence, *Ardipithecus ramidus* or a species closely related to it gave rise to the australopithecines, closer than *Ardipithecus* in overall anatomy to modern humans and better still at bipedal walking. In accordance with this breakthrough, the entire body was refashioned by natural selection to serve an erect posture. The legs were lengthened and straightened, and the feet were elongated to permit an energy-efficient rocking movement during locomotion. The pelvis was sculpted into a shallow bowl to support the viscera. The center of gravity of *Ardipithecus* was now above the legs, instead of at the belly and spine as in the chimpanzees and other anthropoid apes.

With bodies erect and nearly human-like in form, there followed the origin of multiple species among the australopithecines. During the period 3.5 to 2 million years ago, as many as four species of *Australopithecus* (*A. afarensis*, *A. bahrelghazali*, *A. deyiremeda*, *A. platyops*), and the closely related *Kenyanthropus* may have existed together in East and Central Africa. Insofar as can be told from the still fragmentary remains, the australopithecines appear to have been products of what evolutionary biologists call an adaptive radiation. Different degrees of robustness in the teeth and jaws reflect specialization among the competing species in the kind of food they ate. In general, the larger and heavier their teeth and bone relative to the size of their skull, the coarser the vegetation they could include in their diet.

Adaptive radiations that produce closely related species generally reduce competition and allow more to coexist in the same geographical localities. Where they come into contact they tend to diverge from one another in anatomy and behavior, a scattering that reduces competition further. The phenomenon, called character displacement, may have played an important role throughout human evolution.

Understanding the processes of speciation, and with it the follow-up of partial hybridization, character displacement, and adaptive radiation may help explain some of the complex variation encountered in the remains of most of humanity's antecedents. Among them are the earliest members of the genus *Homo*, including *Homo habilis*, the recently discovered fossils at Dmanisi (in the country of Georgia),

Homo georgicus, and the recently discovered species *Homo naledi* of South Africa. Another puzzle that may be solved is the relation in origin and competition among the Neanderthals, the Denisovans, and *Homo sapiens*.

One more principle of evolutionary biology that can help our interpretation of early human evolution is composite evolution. "Missing links" between species, from primitive to more advanced, tend to be a mosaic: some parts of the anatomy are typically more advanced than others. The reason is that different traits tend to evolve at different rates. A striking example comes from the first ant fossils discovered of Mesozoic age in New Jersey deposits, about ninety million years old—twenty-five million years older than the earlier record-holders. These Mesozoic ancestors or near ancestors were a mosaic between the ancestral wasp species and the first ants derived from them. In particular, the mandibles of the fossils are like those of a wasp, the waist and metapleural gland are those of an ant, and the antennae are intermediate between those of wasps and ants. As the first to study them, I anointed the fossils with the scientific name *Sphecomyrma*, which means "wasp ant."

A premier example of mosaic evolution in our own ancestry is provided by *Homo naledi*, a species described in 2015 from abundant fossils found at the Rising Star Cave in South Africa. Elements of the *naledi* body, particularly in the hands, feet, and parts of the skull, are close to modern *Homo*. The brain, however, is about the size of an orange, between 450 and 550 cubic centimeters, a volume closer to

Humanity arose on the African savanna from a line of australopiths by essentially the same route as the other known eusocial animals. A major driving force in social evolution was competition between groups, frequently violent. The final surge to the *Homo* level was enabled by the combination of an initially large brain, fire from the frequent lightning-struck savanna that could be captured and controlled, and the advantages of tightly gathered groups of cooperating members.

chimpanzees than to modern humans, and within the range of our ancestral australopithecines.

Throughout all of humanity's evolutionary prehistory, the most important event was the origin of *Homo habilis* three to two million years ago. The forests were opening, and the savanna combinations of grassland and newly isolated dry-forest copses were spreading. The hominid species, the australopithecines and early *Homo* among them, shifted from a diet based almost exclusively on the C_3 photosynthesis of trees and shrubs, much like that of modern chimpanzees, toward a diet made from C_4 photosynthesis that prevails in grasses, sedges, and succulents typical of tropical savannas and deserts.

The ancestral australopithecine species lived in a primary habitat differing not only in vegetation, but in other basic properties of its ecosystems as well. As the terrain became more open, large animals could be more easily seen and tracked and predators more easily evaded. Cross-country travel was less impeded and more accurate.

Inherent in the savanna environment was another feature even more important for the emergence of humanity: the frequent occurrence of lightning-struck ground fires. With fire that swept through the ground cover during high winds came cooked animal carcasses. Scavengers were likely presented more frequently with bonanzas of meat from animals, including those larger than lizards and mice. Even a small increase in food gathering offered a large benefit. All things considered, meat is the best of foods for creatures limited by caloric intake. It yields higher energy per gram than do fruits and vegetables.

Present-day chimpanzees gather fruit and other vegetable matter while moving in bands within their occupied terrain. They call to one another when a fruiting tree is found. A very small percentage of their calories is obtained by squads of males that cooperate to hunt vervet monkeys, whose raw meat they occasionally share with other members of the larger troop.

Searching through newly burned terrain, a population of australopithecines, perhaps pressured by competing species, turned more toward scavenging of meat as a major addition to their originally primary vegetational diet. An increase in scavenging and predation is enhanced in caloric yield if a defended campsite were established from which scouts and hunters could be dispatched while guards and nurses stayed behind to defend the site and the young gathered there.

In my perception, shared in degree and detail among many anthropologists and biologists, the ecological stage had now been set for the rapid evolutionary growth of the brain. In essence, humans progressed to the level of eusociality by essentially the same route as a few other mammal species, for example African wild dogs. They created nest sites, protected by some of the group, from which others could depart to hunt and forage. Upon return of the hunters and gatherers, the food could be distributed around the entire group. This adaptation led to cooperation and a division of labor based on a relatively high level of social intelligence.

The scenario shared among many scientists is as follows. About a million years ago the controlled use of fire was achieved. Firebrands from lightning strikes carried to other sites bestowed enormous advantages on all aspects of our ancestors' existence. The control of fire improved the yield of meat, allowing more animals to be flushed and trapped. Animals killed by the brush fire were also often cooked by it. And even in the earliest days of the carnivorous *Homo*, the advantage of meat, sinews, and bone made more easily rendered and consumed, had significant consequences. In later evolution, the mastication and physiology of digestion evolved to favor cooked meat and vegetables. Cooking became thereafter a universal human trait. And with cooking came shared meals and a powerful means of social bonding.

Firebrands carried about from one place to another have always been a vital resource, comparable to meat, fruit, and weapons. Tree limbs and bundles of twigs can smolder for hours. With meat, fire, and cooking, campsites could last for more than a few days at a time, persistent enough to be guarded as a refuge. Such a nest, as it can be called in the language of zoology, is the precursor to the attainment of eusociality in all other known animal species. There is evidence of fossil campsites and their accoutrements as far back as *Homo erectus*, the ancestral species intermediate in brain size between *Homo habilis* and modern *Homo sapiens*.

Along with fireside living came division of labor. It was spring-loaded. That is, a predisposition within groups to self-organize into dominance hierarchies already existed. There were in addition existing differences between males and females and between young and old. Further, within each subgroup there were variations in leadership ability as well as in proneness to remain at the campsite. The inevitable result emerging from these preadaptations was, as in all other known eusocial animal species, a complex division of labor.

There followed the fastest evolution on record in the emergence of a complex biological organ. From the australopithecine level of about 400 to 500 cubic centimeters (cc), the cranial capacity rose through the *Homo habilis* grade to 900 cc in *Homo erectus*, the European and Asian direct ancestor of *Homo sapiens*, to the present 1400 cc or more in our species.

The role of group selection in the evolutionary origin of human societies has been primary, although entangled with individual-level selection. To understand what we know about our own origins, or at least think we know, it is useful to return for a moment to the more elementary organization of our phylogenetic cousins, the chimpanzees and bonobos. Their instinctive behavior is overlaid by a thin layer of culture. These great apes of Africa live in communities of up to one hundred fifty members, who join in defending territories, usually by violent means. Each community consists of fluidly changing bands, each typically containing five to ten members. Aggressive behavior is common within both communities and bands, and is most so between bands. Males are usually the chief aggressors at this individual level, and their purpose is to gain status and dominance for themselves and for their bands.

Young males within communities often form gangs that launch border raids, the purpose of which is evidently to kill or drive out members of another community and acquire new territory. The entirety of chimpanzee conquest under natural conditions was witnessed by John Mitani of the University of Michigan and his collaborators in Uganda's Kibale National Park. The war—more precisely, a series of border raids—was conducted over ten years.

The full campaign was eerily human-like. Every ten to fourteen days, patrols of up to twenty males penetrated enemy territory, moving quietly in single file, scanning the terrain from ground to the treetops, and halting cautiously at every close-by noise. If a force larger than their own was encountered, the invaders broke rank and ran back to their own territory. When they encountered a lone male, however, they piled on him in a crowd and bit him to death. When a female was encountered, they usually let her go. This tolerance was no display of gallantry, however. If she carried an infant, they took it from her and killed and ate it. Finally, after pressure applied so long and relentlessly in the Kibale National Park, the residents left and the invading gangs simply annexed the enemy territory, adding 22 percent to the land controlled by their own community.

It has been an entirely reasonable hypothesis among many anthropologists that the border raids and killing among chimpanzees is an incidental outcome of aggression, raised to an abnormal intensity by witnessing human destructiveness, which includes deforestation of the chimpanzee habitats, introduction of disease, and the hunting of chimpanzees for food. Other anthropologists have favored the competing explanation based on evolutionary biology that the chimpanzee depredations are genetically adaptive, having evolved independently of human influence.

In 2014 an international team of thirty anthropologists and biologists compiled all of the well-documented killings by chimpanzees. They found that over 90 percent of the attacks were conducted by males, and two-thirds were between communities as opposed to bands that made up the communities. A great deal of variation occurred in the magnitude of aggression from one community to the next, but it was not correlated with differences in human activity around the chimpanzee populations. It could be seen by direct observation that winners in the border conflicts increased the survival and reproduction of their communities. In other words, chimpanzee war drove group selection.

Lethal violence during warfare is so common in human societies as to suggest that it is an adaptive instinct of our own species. Not only has it been nearly global, but of comparable mortality equal to the between-group war of chimpanzees. Some of the supporting data are given in the accompanying table.

Hunter-gatherer societies, judging from their archaeological remains and the very few that have survived into modern time, provide a window on the origin of humanity as a species. People lived in bands consisting largely of kin. They were linked to other bands by kin ties and marriages. They were loyal to the aggregate of bands as a whole, although never so much as to preclude murder and revenge raids now and then. They tended to be suspicious, fearful, and occasionally hostile to other communities of bands. Lethal violence was a commonplace. The original population of Australia that persisted before colonization provides valuable evidence. Azar Gat, a researcher at Tel Aviv University, has written, "The range of evidence from across Aboriginal Australia, the only continent of hunter-gatherers, strikingly demonstrates that deadly human violence, including group fighting, existed at all social levels, in all population densities, in the simplest of social organization, and in all types of environments."

While in raw combat human tribal aggression resembles that of chimpanzees, it is more complexly organized at the individual level. One of the best illustrations of the refinement in detail has been worked out in the Yanomamö of the northern Amazon basin by Napoleon A. Chagnon and other anthropologists. Violent aggression is territorial in the sense that villages are often in conflict with one another, and as a consequence those with fewer than forty individuals cannot long survive. As individual relationships become more complex, there is blurring of the structure of the kinship groups. Coalitions are commonly formed by individuals of different lineages who live in separate villages. They comprise men

Table 1.

Archaeological and ethnographic evidence on the fraction of adult mortality due to warfare.

"Before present" in the middle heading indicates before 2008.

[From Samuel Bowles, "Did warfare among ancestral hunter-gatherers affect the evolution of human social behaviors?" Science *324(5932): 1295 (2009). Primary references are not included in the table reproduced here.]*

Site	*Archaeological evidence* approx. date (years before present)	*Fraction of adult mortality due to warfare*
British Columbia (30 sites)	5500–334	0.23
Nubia (site 117)	14,000–12,000	0.46
Nubia (near site 117)	14,000–12,000	0.03
Vasiliv'ka III, Ukraine	11,000	0.21
Volos'ke, Ukraine	"Epipalaeolithic"	0.22
S. California (28 sites)	5500–628	0.06
Central California	3500–500	0.05
Sweden (Skateholm 1)	6100	0.07
Central California	2415–1773	0.08
Sarai Nahar Rai, N. India	3140–2854	0.30
Central California (2 sites)	2240–238	0.04
Gobero, Niger	16,000-8200	0.00
Calumnata, Algeria	8300-7300	0.04
Ile Teviec, France	6600	0.12
Bogebakken, Denmark	6300–5800	0.12

Population, region	*Ethnographic evidence* (dates)	*Fraction of adult mortality due to warfare*
Ache, Eastern Paraguay*	Precontact (1970)	0.30
Hiwi, Venezuela–Colombia*	Precontact (1960)	0.17
Murngin, NE Australia*†	1910–1930	0.21
Ayoreo, Bolivia-Paraguay‡	1920–1979	0.15
Tiwi, N. Australia§	1893–1903	0.10
Modoc, N. California§	"Aboriginal times"	0.13
Casiguran Agta, Philippines*	1936–1950	0.05
Anbara, N. Australia*†‖	1950–1960	0.04

* Foragers. † Maritime. ‡ Seasonal forager-horticulturalists.
§ Sedentary hunter-gatherers. ‖ Recently settled.

of similar age and are most commonly maternal cousins. When they kill together, they accrue prestige as a special caste called *unokai*, and they typically come to live in the same village.

This degree of coalition and alliance formation highlights differences in social structure that distinguish humans from chimpanzees and other social primates. But the resulting organization does not diminish the importance of group-level competition as a driving force of human social evolution. On the contrary, it is entirely reasonable that such alliances have been favored throughout human history by cultural evolution. As illustrated by the mathematical models conceived by Maxime Derex and fellow researchers at the

Storytelling among the Ju'hoansi San.

University of Montpelier, group size and cultural complexity are mutually reinforcing in the coevolution of heredity and culture. The larger the group size, the more frequently innovations occur within the group. Communal knowledge deteriorates more slowly, and cultural diversity is preserved more fully and for longer periods.

There is a growing consensus among paleontologists that the origin of our species—and the massive cerebral memory banks that define it—were forged in the firelight of African campsites. The impetus was the cooking of meat, as I've cited here, first by lightning-struck ground fires scavenged by the tribal hunters, and later by firebrands carried from site to site. Cooked meat is a high-energy and very digestible food, easily transported by groups on the move. It led to the clustering of band members and gave advantage to conversation and the division of labor. Cooperative and altruistic behavior in service to the group as a whole were achieved in mental evolution. Social intelligence became premium.

The campsite talk of the early *Homo*, beginning with *habilis*-grade populations, can only be guessed. A general idea of its content can, however, be deduced from conversations within groups of the remaining contemporary hunter-gatherers. Given the importance of this evidence, it is surprising how slow in coming have been careful analyses of the conversations. One, of the Ju/'hoansi (!Kung) of southern Africa and recorded by the anthropologist Polly W. Wiessner, reveals a striking difference between "day talk," centered on food gathering, resource distribution, and other economic matters, and "night talk," primarily devoted to stories, some about living individuals, some enthralling, the latter turning easily into singing, dancing, and religious conversation. At night, the great bulk of the conversation, about 40 percent, consisted of stories and another 40 percent were devoted to myths. In the day, only a very few concerned stories and none involved myths.

> *In the late afternoon, families gathered at their own fires for the evening meal. After dinner and dark, the harsher mood of the day mellowed and people who were in the mood gathered around single fires to talk, make music, or dance. Some nights large groups convened and other nights smaller groups. The focus of conversation changed radically as economic concerns and social gripes were put aside. At this time came 81 percent of lengthy conversations . . .*
>
> *Both men and women told stories, particularly older people who had mastered the art. Camp leaders were frequently good storytellers, although not exclusively so. Two of the best storytellers in the 1970s were blind but cherished for their humor and verbal skills. Stories provided a win-win situation: those who thoroughly engaged others were likely to gain recognition as their stories traveled. Those who listened were entertained while collecting the experiences of others with no direct cost. Because story telling is so important for remembering and knowing people beyond the*

camp, there is likely to have been strong social selection for the manipulation of language to convey characters and emotions.

From the earliest *Homo* formed, as brain size increased, the time devoted to social interactions likely increased. The trend upward has been inferred by Robin I. M. Dunbar of the University of Oxford. He used two correlations from existing species of monkeys and apes: first, time spent grooming as a function of group size, and second, the relation among apes between group size and cranial capacity. Extended to the australopithecines and the *Homo* line of species born from them, this method—admittedly tenuous—suggests that the "required social time" evolved from about one hour a day to two hours in the earliest species of *Homo*, thence four to five hours in modern humanity. In short, longer social interaction is a key component in the evolution of a larger brain and higher intelligence.

REFERENCES AND FURTHER READING

Chapter 1
THE SEARCH FOR GENESIS

Darwin, C. 1859. *On the Origin of Species* (London: John Murray).

Haidt, J. 2012. *The Righteous Mind: Why Good People Are Divided by Politics and Religion* (New York: Pantheon Books).

Ruse, M., and J. Travis, eds. 2009. *Evolution: The First Four Billion Years* (Cambridge, MA: Belknap Press of Harvard University Press).

Standen, E. M., T. Y. Du, and H. C. E. Larsson. 2014. Developmental plasticity and the origin of tetrapods. *Nature* 513(7516): 54–58.

West-Eberhard, M. J. 2003. *Developmental Plasticity and Evolution* (New York: Oxford University Press).

Wilson, E. O. 2014. *The Meaning of Human Existence* (New York: Liveright).

Wilson, E. O. 2015. *The Social Conquest of Earth* (New York: Liveright).

Chapter 2
THE GREAT TRANSITIONS OF EVOLUTION

An, J. H., E. Goo, H. Kim, Y-S. Seo, and I. Hwang. 2014. Bacterial quorum sensing and metabolic slowing in a cooperative population. *Proceedings of the National Academy of Sciences, USA* 111(41): 14912–14917.

Maynard Smith, J., and E. Szathmáry. 1995. *The Major Transitions of Evolution* (New York: W. H. Freeman Spektrum).

Miller, M. B., and B. L. Bassler. 2001. Quorum sensing in bacteria. *Annual Review of Microbiology* 55: 165-199.

Wilson, E. O. 1971. *The Insect Societies* (Cambridge, MA: Belknap Press of Harvard University Press).

Chapter 3
THE GREAT TRANSITIONS DILEMMA AND HOW IT WAS SOLVED

Boehm, C. 2012. *Moral Origins: The Evolution of Virtue, Altruism, and Shame* (New York: Basic Books).

Graziano, M. S. N. 2013. *Consciousness and the Social Brain* (New York: Oxford University Press).

Haidt, J. 2012. *The Righteous Mind: Why Good People Are Divided by Politics and Religion* (New York: Pantheon Books).

Li, L., H. Peng, J. Kurths, Y. Yang, and H. J. Schellnhuber. 2014. Chaos-order transition in foraging behavior of ants. *Proceedings of the National Academy of Sciences, USA* 111(23): 8392–8397.

Pruitt, J. N. 2013. A real-time eco-evolutionary dead-end strategy is mediated by the traits of lineage progenitors and interactions with colony invaders. *Ecology Letters* 16: 879–886.

Ruse, M., ed. 2009. *Philosophy After Darwin* (Princeton, NJ: Princeton University Press).

Wilson, E. O. 2014. *The Meaning of Human Existence* (New York: Liveright).

Wright, C. M., C. T. Holbrook, and J. N. Pruitt. 2014. Animal personality aligns task specialization and task proficiency in a spider society. *Proceedings of the National Academy of Sciences, USA* 111(26): 9533–9537.

Chapter 4
TRACKING SOCIAL EVOLUTION THROUGH THE AGES

Darwin, C. 1859. *On the Origin of Species* (London: John Murray).

Dunlap, A. S., and D. W. Stephens. 2014. Experimental evolution of prepared learning. *Proceedings of the National Academy of Sciences, USA* 11(32): 11750–11755.

Hendrickson, H., and P. B. Rainey. 2012. How the unicorn got its horn. *Nature* 489(7417): 504–505.

Hutchinson, J. 2014. Dynasty of the plastic fish. *Nature* 513(7516): 37–38.

Maynard Smith, J., and E. Szathmáry. 1995. *The Major Transitions in Evolution* (New York: W. H. Freeman Spektrum).

Melo, D., and G. Marroig. 2015. Directional selection can drive the evolution of modularity in complex traits. *Proceedings of the National Academy of Sciences, USA* 112(2): 470–475.

Standen, E. M., T. Y. Du, and H. C. E. Larsson. 2014. Developmental plasticity and the origin of tetrapods. *Nature* 513(7516): 54–58.

West-Eberhard, M. J. 2003. *Developmental Plasticity and Evolution* (New York: Oxford University Press).

Chapter 5
THE FINAL STEPS TO EUSOCIALITY

Bang, A., and R. Gadagkar. 2012. Reproductive queue without overt conflict in the primitively eusocial wasp *Ropalidia marginata*. *Proceedings of the National Academy of Sciences, USA* 109(36): 14494–14499.

Biedermann, P. H. W., and M. Taborsky. 2011. Larval helpers and age polyethism in ambrosia beetles. *Proceedings of the National Academy of Sciences, USA* 108(41): 17064–17069.

Cockburn, A. 1998. Evolution of helping in cooperatively breeding birds. *Annual Review of Ecology, Evolution, and Systematics* 29: 141–177.

Costa, J. T. 2006. *The Other Insect Societies* (Cambridge, MA: Belknap Press of Harvard University Press).

Derex, M., M.-P. Beugin, B. Godelle, and M. Raymond. 2013. Experimental evidence for the influence of group size on cultural complexity. *Nature* 503(7476): 389–391.

Evans, H. E. 1958. The evolution of social life in wasps. *Proceedings of the Tenth International Congress of Entomology* 2: 449–451.

Hölldobler, B., and E. O. Wilson. *The Ants* (Cambridge, MA: Belknap Press of Harvard University Press).

Hunt, J. H. 2011. A conceptual model for the origin of worker behaviour and adaptation of eusociality. *Journal of Evolutionary Biology* 25: 1–19.

Liu, J., R. Martinez-Corral, A. Prindle, D.-Y. D. Lee, J. Larkin, M. Gabalda-Sagarra, J. Garcia-Ojalvo, and G. M. Süel. 2017.

Coupling between distant biofilms and emergence of nutrient time-sharing. *Science* 356(6338): 638–642.

Michener, C. D. 1958. The evolution of social life in bees. *Proceedings of the Tenth International Congress of Entomology* 2: 441–447.

Nalepa, C. A. 2015. Origin of termite eusociality: Trophallaxis integrates the social, nutritional, and microbial environment. *Ecological Entomology* 40(4): 323–335.

Pruitt, J. N. 2012. Behavioural traits of colony founders affect the life history of their colonies. *Ecology Letters* 15: 1026–1032.

Rendueles, O., P. C. Zee, I. Dinkelacker, M. Amherd, S. Wielgoss, and G. J. Velicer. 2015. Rapid and widespread de novo evolution of kin discrimination. *Proceedings of the National Academy of Sciences, USA* 112(29): 9076–9081.

Richerson, P. 2013. Group size determines cultural complexity. *Nature* 503(7476): 351–352.

Rosenthal, S. B., C. R. Twomey, A. T. Hartnett, H. S. Wu, and I. D. Couzin. 2015. Revealing the hidden networks of interaction in mobile animal groups allows prediction of complex behavioral contagion. *Proceedings of the National Academy of Sciences, USA* 112(15): 4690–4695.

Szathmáry, E. 2011. To group or not to group? *Science* 334(6063): 1648–1649.

Wilson, E. O. 1971. *The Insect Societies* (Cambridge, MA: Belknap Press of Harvard University Press).

Wilson, E. O. 1975. *Sociobiology: The New Synthesis* (Cambridge, MA: Belknap Press of Harvard University Press).

Wilson, E. O. 1978. *On Human Nature* (Cambridge, MA: Harvard University Press).

Wilson, E. O. 2008. One giant leap: How insects achieved altruism and colonial life. *BioScience* 58(1): 17–25.

Wilson, E. O., and M. A. Nowak. 2014. Natural selection drives the evolution of ant life cycles. *Proceedings of the National Academy of Sciences, USA* 111(35): 12585–12590.

Chapter 6
GROUP SELECTION

Abbot, P., J. H. Withgott, and N. A. Moran. 2001. Genetic conflict and conditional altruism in social aphid colonies. *Proceedings of the National Academy of Sciences, USA* 98(21): 12068–12071.

Abouheif, E., and G. A. Wray. 2002. Evolution of the gene network underlying wing polyphenism in ants. *Science* 297(5579): 249–252.

Adams, E. S., and M. T. Balas. 1999. Worker discrimination among queens in newly founded colonies of the fire ant *Solenopsis invicta*. *Behavioral Ecology and Sociobiology* 45(5): 330–338.

Allen, B., M. A. Nowak, and E. O. Wilson. 2013. Limitations of inclusive fitness. *Proceedings of the National Academy of Sciences, USA* 110(50): 20135–20139.

Avila, P., and L. Fromhage. 2015. No synergy needed: Ecological constraints favor the evolution of eusociality. *American Naturalist* 186(1): 31–40.

Bang, A., and R. Gadagkar. 2012. Reproductive queue without overt conflict in the primitively eusocial wasp *Ropalidia marginata*. *Proceedings of the National Academy of Sciences, USA* 109(36): 14494–14499.

Birch, J., and S. Okasha. 2015. Kin selection and its critics. *BioScience* 65(1): 22–32.

Boehm, C. 2012. *Moral Origins: The Evolution of Virtue, Altruism, and Shame* (New York: Basic Books).

Bourke, A. F. G. 2013. A social rearrangement: Genes and queens. *Nature* 493(7434): 612.

De Vladar, H. P., and E. Szathmáry. 2017. Beyond Hamilton's rule. *Science* 356(6337): 485–486.

Gat, A. 2018. Long childhood, family networks, and cultural exclusivity: Missing links in the debate over human group selection and altruism. *Evolutionary Studies in Imaginative Culture* 2(1): 49–58.

Haidt, J. 2012. *The Righteous Mind: Why Good People Are Divided by Politics and Religion* (New York: Pantheon Books).

Hölldobler, B., and E. O. Wilson. 2009. *The Superorganism: The Beauty, Elegance, and Strangeness of Insect Societies* (New York: W. W. Norton).

Hunt, J. H. 2018. An origin of eusociality without kin selection. In preparation.

Kapheim, K. M., et al. 2015. Genomic signatures of evolutionary transitions from solitary to group living. *Science* 348(6239): 1139–1142.

Linksvayer, T. 2014. Evolutionary biology: Survival of the fittest group. *Nature* 514(7522): 308–309.

Mank, J. E. 2013. A social rearrangement: Chromosome mysteries. *Nature* 493(7434): 612–613.

Nalepa, C. I. 2015. Origin of termite eusociality: Trophallaxis integrates the social, nutritional, and microbial environments. *Ecological Entomology* 40(4): 323–335.

Nowak, M. A., A. McAvoy, B. Allen, and E. O. Wilson. 2017. The general form of Hamilton's rule makes no predictions and cannot be tested empirically. *Proceedings of the National Academy of Sciences, USA* 114(22): 5665–5670.

Oster, G. F., and E. O. Wilson. 1978. *Caste and Ecology in the Social Insects* (Princeton, NJ: Princeton University Press).

Pruitt, J. N. 2012. Behavioural traits of colony founders affect the life history of their colonies. *Ecology Letters* 15: 1026–1032.

Pruitt, J. N. 2013. A real-time eco-evolutionary dead-end strategy is mediated by the traits of lineage progenitors and interactions with colony invaders. *Ecology Letters* 16: 879–886.

Pruitt, J. N., and C. J. Goodnight. 2014. Site-specific group selection drives locally adapted group compositions. *Nature* 514(7522): 359–362.

Rendueles, O., P. C. Zee, I. Dinkelacker, M. Amherd, S. Wielgoss, and G. J. Velicer. 2015. Rapid and widespread de novo evolution of kin discrimination. *Proceedings of the National Academy of Sciences, USA* 112(29): 9076–9081.

Ruse, M., and J. Travis, eds. 2009. *Evolution: The First Four Billion Years* (Cambridge, MA: Belknap Press of Harvard University Press).

Science and Technology: Ecology. 2015. Pack power. *The Economist*, 30 May: 79–80.

Shbailat, S. J., and E. Abouheif. 2013. The wing patterning network in the wingless castes of myrmicine and formicine species is a mix of evolutionarily labile and non-labile genes. *Journal of Experimental Zoology (Part B: Molecular and Developmental Evolution)* 320: 74–83.

Silk, J. B. 2014. Animal behaviour: The evolutionary roots of lethal conflict. *Nature* 513(7518): 321–322.

Teseo, S., D. J. Kronauer, P. Jaisson, and N. Châline. 2013. Enforcement of reproductive synchrony via policing in a clonal ant. *Current Biology* 23(4): 328–332.

Thompson, F. J., M. A. Cant, H. H. Marshall, E. I. K. Vitikainen, J. L. Sanderson, H. J. Nichols, J. S. Gilchrist, M. B. V. Bell, A. J. Young, S. J. Hodge, and R. A. Johnstone. 2017. Explaining negative kin discrimination in a cooperative mammal society. Proceedings of the National Academy of Sciences, USA 114(20): 5207–5212.

Tschinkel, W. R. 2006. *The Fire Ants* (Cambridge, MA: Belknap Press of Harvard University Press).

Wang, J., Y. Wurm, M. Nipitwattanaphon, O. Riba-Grognuz, Y.-C. Huang, D. Shoemaker, and L. Keller. 2013. A Y-like social chromosome causes alternative colony organization in fire ants. *Nature* 493(7434): 664–668.

Wilson, D. S., and E. O. Wilson. 2007. Rethinking the theoretical foundation of sociobiology. *Quarterly Review of Biology* 82(4): 327–348.

Wilson, E. O. 1971. *The Insect Societies* (Cambridge, MA: Harvard University Press).

Wilson, E. O. 2008. One giant leap: How insects achieved altruism and colonial life. *BioScience* 58(1): 17–24.

Wilson, E. O. 2012. *The Social Conquest of Earth* (New York: Liveright).

Wilson, M. L., et al. 2014. Lethal aggression in *Pan* is better explained by adaptive strategies than human impacts. *Nature* 513(7518): 414–417.

Wright, C. M., C. T. Holbrook, and J. N. Pruitt. 2014. Animal personality aligns task specialization and task proficiency in a spider society. *Proceedings of the National Academy of Sciences, USA* 111(26): 9533–9537.

Chapter 7
THE HUMAN STORY

Aanen, D. K., and T. Blisseling. 2014. The birth of cooperation. *Science* 345(6192): 29–30.

An, J. H., E. Goo, H. Kim, Y.-S. Seo, and I. Hwang. 2014. Bacterial quorum sensing and metabolic slowing in a cooperative population. *Proceedings of the National Academy of Sciences, USA* 111(41): 14912–14917.

Antón, S. C., R. Potts, and L. C. Aiello. 2014. Evolution of early *Homo*: An integrated biological perspective. *Science* 345(6192): 45.

Barragan, R. C., and C. S. Dweck. 2014. Rethinking natural altruism: Simple reciprocal interactions trigger children's benevolence. *Proceedings of the National Academy of Sciences, USA* 111(48): 17071–17074.

Bateman, T. S., and A. M. Hess. 2015. Different personal propensities among scientists relate to deeper vs. broader knowledge contributions. *Proceedings of the National Academy of Sciences, USA* 112(12): 3653–3658.

Boardman, J. D., B. W. Domingue, and J. M. Fletcher. 2012. How social and genetic factors predict friendship networks. *Proceedings of the National Academy of Sciences, USA* 109(43): 17377–17381.

Boehm, C. 2012. *Moral Origins: The Evolution of Virtue, Altruism, and Shame* (New York: Basic Books).

Botero, C. A., B. Gardner, K. R. Kirby, J. Bulbulia, M. C. Gavin, and R. D. Gray. 2014. The ecology of religious beliefs. *Proceedings of the National Academy of Sciences, USA* 111(47): 16784–16789.

Brown, K. S., C. W. Marean, Z. Jacobs, B. J. Schoville, S. Oestmo, E. C. Fisher, J. Bernatchez, P. Karkanas, and T. Matthews. 2012. An early and enduring advanced technology originating 71,000 years ago in South Africa. *Nature* 491(7425): 590–593.

Cockburn, A. 1998. Evolution of helping in cooperatively breeding birds. *Annual Review of Ecology, Evolution, and Systematics* 29: 141–177.

Crockett, M. J., Z. Kurth-Nelson, J. Z. Siegel, P. Dayan, and R. J. Dolan. 2014. Harm to others outweighs harm to self in moral decision making. *Proceedings of the National Academy of Sciences, USA* 111(48): 17320–17325.

Di Cesare, G., C. Di Dio, M. Marchi, and G. Rizzolatti. 2015. Expressing our internal states and understanding those of others. *Proceedings of the National Academy of Sciences, USA* 112(33): 10331–10335.

Dunbar, R. I. M. 2014. How conversations around campfires came to be. *Proceedings of the National Academy of Sciences, USA* 111(39): 14013–14014.

Flannery, K. V., and J. Marcus. 2012. *The Creation of Inequality: How Our Prehistoric Ancestors Set the Stage for Monarchy, Slavery, and Empire* (Cambridge, MA: Harvard University Press).

Foer, J. 2015. It's time for a conversation (dolphin intelligence). *National Geographic* 227(5): 30–55.

Gallo, E., and C. Yan. 2015. The effects of reputational and social knowledge on cooperation. *Proceedings of the National Academy of Sciences, USA* 112(12): 3647–3652.

Gintis, H. 2016. *Individuality and Entanglement: The Moral and Material Bases of Social Life* (Princeton, NJ: Princeton University Press).

Gómez, J. M., M. Verdu, A. González-Megías, and M. Méndez. 2016. The phylogenetic roots of human lethal violence. *Nature* 538(7624): 233–237.

González-Forero, M., and S. Gavrileta. 2013. Evolution of manipulated behavior. *American Naturalist* 182(4): 439–451.

Gottschall, J., and D. S. Wilson, eds. 2005. *The Literary Animal:*

Evolution and the Nature of Narrative (Evanston, IL: Northwestern University Press).

Halevy, N., and E. Halali. 2015. Selfish third parties act as peacemakers by transforming conflicts and promoting cooperation. *Proceedings of the National Academy of Sciences, USA* 112(22): 6937–6942.

Heinrich, B. 2001. *Racing the Antelope: What Animals Can Teach Us About Running and Life* (New York: Cliff Street).

Hilbe, C., B. Wu, A. Traulsen, and M. A. Nowak. 2014. Cooperation and control in multiplayer social dilemmas. *Proceedings of the National Academy of Sciences, USA* 111(46): 16425–16430.

Hoffman, M., E. Yoeli, and M. A. Nowak. 2015. Cooperate without looking: Why we care what people think and not just what they do. *Proceedings of the National Academy of Sciences, USA* 112(6): 1727–1732.

Keiser, C. N., and J. N. Pruitt. 2014. Personality composition is more important than group size in determining collective foraging behaviour in the wild. *Proceedings of the Royal Society B* 281(1796): 1424–1430.

Leadbeater, E., J. M. Carruthers, J. P. Green, N. S. Rosen, J. Field. 2011. Nest inheritance is the missing source of direct fitness in a primitively eusocial insect. *Science* 333(6044): 874–876.

LeBlanc, S. A., and K. E. Register. 2003. *Constant Battles: The Myth of the Peaceful, Noble Savage* (New York: St. Martin's Press).

Liu, J., R. Martinez-Corral, A. Prindle, D.-Y. D. Lee, J. Larkin, M. Gabalda-Sagarra, J. Garcia-Ojalvo, and G. M. Süel. 2017. Coupling between distant biofilms and emergence of nutrient time-sharing. *Science* 356(6338): 638–642.

Macfarlan, S. J., R. S. Walker, M. V. Flinn, and N. A. Chagnon. 2014. Lethal coalitionary aggression and long-term alliance formation among Yanomamö men. *Proceedings of the National Academy of Sciences, USA* 111(47): 16662–16669.

Martinez, A. E., and J. P. Gomez. 2013. Are mixed-species bird flocks stable through two decades? *American Naturalist* 181(3): E53–E59.

Mesterton-Gibbons, M., and S. M. Heap. 2014. Variation between self- and mutual assessment in animal contests. *American Naturalist* 183(2): 199–213.

Miller, M. B., and B. L. Bassler. 2001. Quorum sensing in bacteria. *Annual Review of Microbiology* 55: 165–199.

Muchnik, L., S. Aral, and S. J. Taylor. 2013. Social influence bias: A randomized experiment. *Science* 341(6146): 647–651.

Opie, C., et al. 2014. Phylogenetic reconstruction of Bantu kinship challenges main sequence theory of human social evolution. *Proceedings of the National Academy of Sciences, USA* 111(49): 17414–17419.

Rand, D. G., M. A. Nowak, J. H. Fowler, and N. A. Christakis. 2014. Static network structure can stabilize human cooperation. *Proceedings of the National Academy of Sciences, USA* 111(48): 17093–17098.

Roes, F. L. 2014. Permanent group membership. *Biological Theory* 9(3): 318–324.

Suderman, R., J. A. Bachman, A. Smith, P. K. Sorger, and E. J. Deeds. 2017. Fundamental trade-offs between information flow in single cells and cellular populations. *Proceedings of the National Academy of Sciences, USA* 114(22): 5755–5760.

Thomas, E. M. 2006. *The Old Way: A Story of the First People* (New York: Farrar, Straus and Giroux).

Tomasello, M. 1999. *The Cultural Origins of Human Cognition* (Cambridge, MA: Harvard University Press).

Wiessner, P. W. 2014. Embers of society: Firelight talk among the

Ju/'hoansi Bushmen. *Proceedings of the National Academy of Sciences, USA* 111(39): 14027–14035.

Wilson, E. O. 1975. *Sociobiology: The New Synthesis* (Cambridge, MA: Belknap Press of Harvard University Press), p. 39.

Wilson, E. O. 2012. *The Social Conquest of Earth* (New York: Liveright).

Wilson, E. O. 2014. *The Meaning of Human Existence* (New York: Liveright).

Wilson, M. L., et al. 2014. Lethal aggression in *Pan* is better explained by adaptive strategies than human impacts. *Nature* 513(7518): 414–417.

Wrangham, R. W. 2009. *Catching Fire: How Cooking Made Us Human* (New York: Basic Books).

Wrangham, R. W., and D. Peterson. 1996. *Demonic Males: Apes and the Origins of Human Violence* (Boston: Houghton Mifflin).

ACKNOWLEDGMENTS

I am grateful for the contributions of many in the composing of the text, and especially to Kathleen M. Horton of Harvard University, and Robert Weil of Liveright Publishing Corporation, for their advice and support; and to James T. Costa, for his vital synthesis of the subsocial steps of arthropods leading to the eusocial stage of evolution.

INDEX

Page numbers in *italics* refer to illustrations